Arduino Servo Projects

Robert J Davis II

Arduino Servo Projects

Copyright 2015 by Robert J Davis II

This is essentially a continuation of my earlier book "Arduino Robotics Projects". These are some more advanced robotics projects and they will use a lot of servos. Some readers have complained that the projects in the earlier book required specific toys in order to work. That was not the intention. Instead the book was meant to teach the basics for robotics and then to show some simple examples showing how to apply what you have learned. This book will repeat some of what was covered in that book in case you did not read it first.

Disclaimer: Once again the safe construction and operation of these devices is solely up to the reader. The reader must take all safety precautions and accept all responsibility for the safe operation of these devices. There are no guarantees implied with the circuit designs and software programs that are found within this book.

Disclaimer 2: No two servos work exactly the same. No two robots are built exactly the same. The software in this book may not work with your robot without some adjustment to the servo positions in software and hardware. This is normal and should be expected.

The most important thing is for you to have lots of fun! Try out some projects in this book and see what you like. Make your own hardware and software improvements. I am sure you can come up with some better designs!

A common acronym that is used when referring to robots is "DOF" or "Degrees of Freedom". It generally refers to the amount of servo motors that are used to give the robot mobility.

Table of Contents

Chapter 1

Working with Servos

Servos are motors that have a built in motor controller and a "feedback" loop. Basically a variable resistor or other device monitors the motor's position. This information is then "fed back" to the built in motor controller. The motor controller takes commands that are usually in the form of a pulse width. Then the controller matches up the motor position with what position the pulse width told it to move the motor to.

Servo's typically come in several sizes, as can be seen in the above picture. There are the larger servos on the left, normal "standard" sized servos in the middle and a micro sized servo on the right. The middle size of servo is the one we will use for most of these projects. You might note that most micro servos have only one mounting hole at each end instead of two mounting holes.

In the case of pulse width modulation commands, usually a pulse width of one millisecond tells the servo to move to zero degrees. A pulse width of 1.5 milliseconds tells the servo to move to 90 degrees. A pulse width of two milliseconds tells the servo to move almost completely open or 180 degrees. The servo "home" position is usually at 90 degrees or in the middle of its range. For some servos the position and pulse width may vary. You might find that for some servos a .5 millisecond pulse results in zero degrees of rotation and a 2.5 millisecond pulse results in 180 degrees of rotation. In any case there is also a 20 millisecond delay between each of the control pulses.

Here is a chart showing the pulse width and the corresponding position of the servo motor.

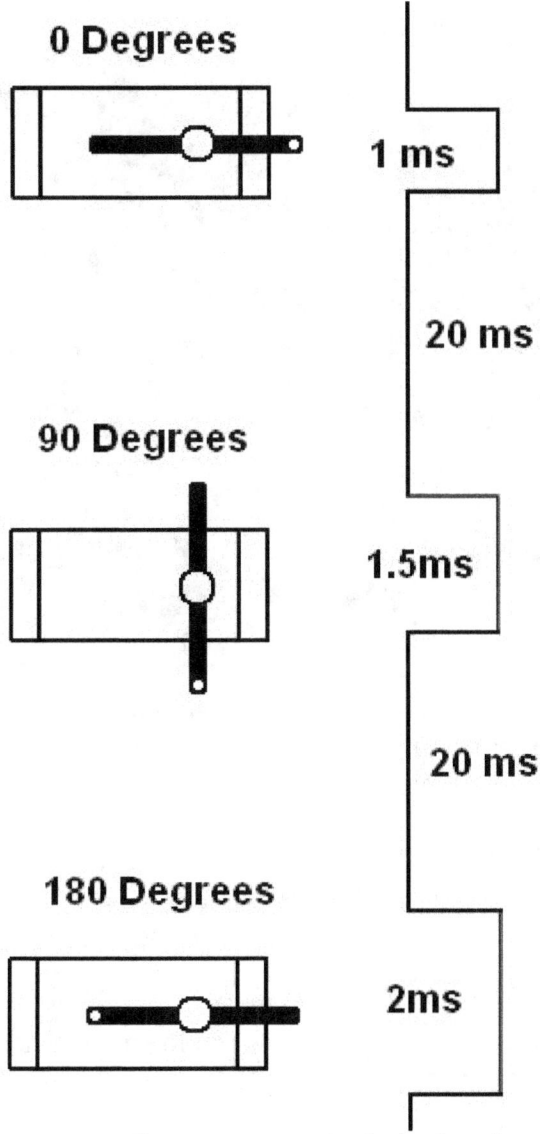

Some of the advantages of servo motors include that the power source does not have to be switched on or off or otherwise controlled. The power to the servo motor can always be left "on". This time a PWM output pin of the Arduino can directly control the servo, no driver circuit or transistor is needed, because there is a driver inside of the servo. Servos make great proportional valve controllers because you can vary a valve from off to full on. For instance, if you want to water your plants automatically and you want the rate of water flow to be adjusted according to the humidity, it can be done with a servo.

Here is a picture of the Arduino servo motor test setup. That is a micro servo being used for the servo demo.

Here is a sketch to demonstrate the operation of a servo motor. For this experiment the center or output of a variable resistor is connected to A0. This demo uses the Arduino servo library.

```
/********************************
// Servo motor demonstration program
// By Bob Davis
// July 10, 2013
// Servo Connected to Gnd, Vin, and D9
// Variable resistor on AO, high end=5V and low end=Gnd
/********************************/
#include <Servo.h>
Servo demoservo;
// The variable resistor is on A0
```

```
int vrpin = 0;
int pos;

void setup() {
// The servo is on pin 9
  demoservo.attach(9);
}

void loop() {
// Read the variable resistor, 1024/5=205 degrees rotation
// Values over 180 are ignored
  pos = analogRead(vrpin)/5;
// send the position to the servo
  demoservo.write(pos);
  delay(25);
}
```

For the projects in this book you will also need about 9 to18 servo motors.
Some of my servos are seen in the next picture. Here are a few of the
needed servos, they are mostly MG995's and MG996's. These are not the
best servos as far as smooth movements, accurate returns, etc. However
these servos will work for most of the projects found in this book.

These are some specs for some popular servos. Basically you want a
normal sized servo with over 100 oz-inches of torque that does not cost a
fortune. Servos with 150 oz-inches of torque would be even better.

Make	Model	Size	Torque 5V	Price
Futaba	S148	Standard	33 oz-in	$15
Futaba	S3001	Standard	33 oz-in	$25
Futaba	S3003	Standard	44 oz-in	$12
Futaba	S3004	Standard	44 oz-in	$13
Futaba	S3010	Standard	72 oz-in	$25
Futaba	S3305	Standard	99 oz-in	$32
Futaba	S9451	Standard	97 oz-in	$70
Futaba	S9650	Standard	50 oz-in	$60
Futaba	BLS451	Standard	118 oz-in	$99
Hitec	HS-311	Standard	42 oz-in	$13
Hitec	HS-322/325	Standard	42 oz-in	
Hitec	HS-422/425	Standard	46 oz-in	
Hitec	HS-475	Standard	61 oz-in	$28
Hitec	HS-625	Standard	76 oz-in	
Hitec	HS-635	Standard	69 oz-in	
Hitec	HS-645	Standard	107 oz-in	$40
Hitec	HS-965	Standard	111 oz-in	
Hitec	HS-985	Standard	144 oz-in	
Hitec	HS-5485	Standard	72 oz-in	
Hitec	HS-5625	Standard	110 oz-in	
Hitec	HS-5645	Standard	143 oz-in	
Hitec	HS-5955	Standard	250 oz-in	$99
TowerPro	SG90/91	Micro	25 oz-in	$2
TowerPro	MG90/91	Micro	30 oz-in	$3
TowerPro	MG930	Standard	50 oz-in	$10
TowerPro	MG945	Standard	138 oz-in	$10
TowerPro	MG946	Standard	145 oz-in	$18
TowerPro	MG955	Standard	97 oz-in	$20
TowerPro	MG956	Standard	104 oz-in	
TowerPro	MG958/968	Standard	250 oz-in	$12
TowerPro	MG995/996	Standard	130 oz-in	$10
Traxxas	2018	Standard	72 (6V)	
Traxxas	2055	Standard	86 (6V)	
Traxxas	2056	Standard	80 (6V)	
Traxxas	2070/75	Standard	125 (6V)	

The typical "Standard" servo size is 1.6×0.8×1.7 inches.

Here is some more Information that I have discovered about servos and robots. Be sure that the servo you buy is not a cheap imitation. This can be detected by the seller hiding the manufacturer name or misspelling the manufacturer name. A common example would be "Tower Pro" (real) or "Towerd Pro" (fake).

This is my new "rule of thumb" for making humanoid, as in two legged robots. The problem is that the more the servos (and their connecting hardware) the more the robot weighs. The leg servos of a 5-9 DOF Humanoid robot should handle 100 oz. inches of torque. The leg servos of a 10-15 DOF Humanoid robot should handle 150 oz. inches of torque. The leg servos of a 16-20 DOF Humanoid robot should handle 200 oz. inches of torque.

One of the things you will need to do before you start assembling anything is to set the servos to 90 degrees. You can force the servo to the ends of its movement and then find the middle but that is not as accurate as using software to find the 90 degree mark. I use a program that will set up to 12 servos attached to D2 to D13 to their 90 degree position or the center of their rotation.

There are an odd number of teeth on the servo output gear so that there are four different possible gear alignments. That is to say there are four chances that the servo will line up with whatever you are connecting it to. If it does not work properly, turn the flange 90 degrees and try again.

Remember that the shoulders for the humanoid robots should be in the middle of their movement so the arms should be straight out or ½ their way up.

```
// Arduino 12 Servo to a "home" of 90 degrees
// Used to set the zero position for all 12 servos
// Mapped as D2 is servo1
// Mapped as D3 is servo2
// Etc.

#include <Servo.h>
Servo servo1; // Define our Servo's
Servo servo2;
Servo servo3;
Servo servo4;
Servo servo5;
```

```cpp
Servo servo6;
Servo servo7;
Servo servo8;
Servo servo9;
Servo servo10;
Servo servo11;
Servo servo12;

void setup(){
  servo1.attach(2); // servo on digital pin 2
  servo2.attach(3); // servo on digital pin 3
  servo3.attach(4); // servo on digital pin 4
  servo4.attach(5); // servo on digital pin 5
  servo5.attach(6); // servo on digital pin 6
  servo6.attach(7); // servo on digital pin 7
  servo7.attach(8); // servo on digital pin 8
  servo8.attach(9); // servo on digital pin 9
  servo9.attach(10); // servo on digital pin 10
  servo10.attach(11); // servo on digital pin 11
  servo11.attach(12); // servo on digital pin 12
  servo12.attach(13); // servo on digital pin 13
}

void loop(){
  servo1.write(90);   // trim to level
  servo2.write(90);
  servo3.write(90);
  servo4.write(90);
  servo5.write(90);
  servo6.write(90);
  servo7.write(90);
  servo8.write(90);
  servo9.write(90);
  servo10.write(90);
  servo11.write(90);
  servo12.write(90);
  delay(1000);        // Wait 1 second
}
// End of program
```

There are many possible solutions to the 12 servo limit. They involve the use of either software or hardware. We will concentrate on the software modifications.

Software Solutions:
   – Modify servo.h
   – Switch to softwareservo.
   – Write your own servo driver.
Hardware Solutions (Not covered here):
   – Switch to an Arduino Mega.
   – Use an external 4017 – 10 channel expansion.
   – Use an external 16 channel servo controller.
   – Use an external 32 channel servo controller.

One solution is to modify servo.h to support more than 12 servos per timer. Basically you find Arduino-1.0.5/libraries/Servo. In that folder there is a file called servo.h, open it with Wordpad. Do not allow wordpad to convert it to a formatted document it must remain a "text document". There is a line in servo.h that says "SERVOS_PER_TIMER". Change the number from 12 to 18 for a total number of 18 servos. Save your changes.

Up next is a picture highlighting the needed modification to servo.h to support more than 12 servos.

```
#define Servo_VERSION           2         // software version of
this library

#define MIN_PULSE_WIDTH         544       // the shortest pulse sent
to a servo
#define MAX_PULSE_WIDTH         2400      // the longest pulse sent
to a servo
#define DEFAULT_PULSE_WIDTH     1500      // default pulse width when
servo is attached
#define REFRESH_INTERVAL        20000     // minumim time to refresh
servos in microseconds

#define SERVOS_PER_TIMER        16        // the maximum number of
servos controlled by one timer
#define MAX_SERVOS    (_Nbr_16timers   * SERVOS_PER_TIMER)

#define INVALID_SERVO           255       // flag indicating an
invalid servo index
```

So you do not want to modify servo.h to support more than 12 servos? The next option is softwareservo. But guess what? It does not work

without modification either! You need to change the line that says
"<Wprogram.h>" to say "<Arduino.h>" using wordpad.

```
#ifndef SoftwareServo_h
#define SoftwareServo_h

#include <WProgram.h>
#include <inttypes.h>

class SoftwareServo
{
  private:
    uint8_t pin;
    uint8_t angle;        // in degrees
    uint16_t pulse0;      // pulse width in TCNT0 counts
    uint8_t min16;        // minimum pulse, 16uS units  (default is 34)
    uint8_t max16;        // maximum pulse, 16uS units, 0-4ms range (del
    class SoftwareServo *next;
    static SoftwareServo* first;
  public:
    SoftwareServo():
```

A third software option is to use your own timing. It is actually not that
difficult to do. Here is a sketch that resets all servos to 90 degrees. All
you need to do is to pulse each servo to the position that you want it to go
to. With a little modification you could have this running your robot!

```
// Servo with manual timing
// By Bob Davis
// November 2, 2015
// Servos Connected to D2-D19
// Does not use servo.h as the timing is included

void setup() {
// The servos on pins 0-19
  for (int i=0; i<20; i++){
  pinMode (I, OUTPUT);
  }
}
void loop() {
  for (int i=0; i<20; i++){
  digitalWrite (I, HIGH);
  delayMicroseconds(1500);
  digitalWrite (I, LOW);
  }
  delay(20); // in milliseconds
} // End of program
```

Chapter 2

The Servo Building Blocks

The projects in this book use what you might call "Servo Building Blocks". These metal parts make it easy to build just about anything using standard sized servos. These projects will take lots of servos. I used a "17 DOF" kit to get most of the parts but you can use the parts list to buy the individual parts. The brackets come in black, white, silver and I have even seen some in blue.

Here are some of the many parts that will come with the walking humanoid robot kit that is found on eBay. At the top is the "U" Shaped brackets, six short ones and four long ones. Below that is the 12 servo motor mounting brackets.

This next picture shows some of the other brackets. At the top there are four straight and four angled connecting brackets. Below that there are the two foot bottoms and the four angled "U" Shaped brackets.

You will also need some servo flanges, some ball bearings, and lots of screws like the ones that are shown in the picture below.

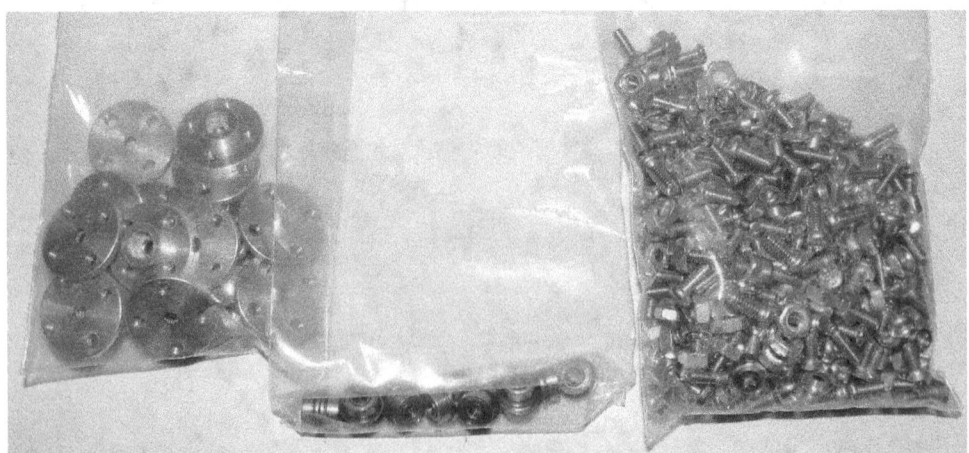

Here is the actual parts list that came with my "17 DOF" robot kit. These are not enough parts to build all of the projects in this book. They will be enough parts for the 9 DOF, 13 DOF, and some forms of the 17 DOF humanoid robot.

- 16 x Multi-functional servo mounting bracket
- 7 x Short U-type servo bracket
- 4 x Long U-type servo bracket
- 4 x Angled U-type servo bracket
- 4 x Flat servo bracket
- 4 x L-type servo bracket
- 1 x Robot waist bracket – or make your own instead
- 2 x Foot Base
- 14 x Miniature Ball Radial Bearing
- 17 x Metal Servo Horn Flange
- 1 x Screw and nut set containing the following parts;
        (64 Larger screws and nuts – Four per servo)
        (8 Larger flat head screws and nuts – Four per foot)
        (128 Smaller screws and nuts – Eight per servo)

These parts are needed but are not included in the robot kit:

- 17 x MG995, MG996 or MG958 Servo's.
  These must be purchased separately for these projects
  For the 17 DOF humanoid project you might want to use some more powerful servos for the hips such as MG958's.
- Arduino Uno and Servo/Sensor Shield
  You can make your own servo shield.

These additional parts are needed for some of the projects in this book:

- 1 x Multi-functional servo mounting bracket
- 3 x Short U-type servo bracket (Dog and Dino Tail)
- 2 x Long U-type servo bracket (Dino Tail)
- 2 x L-type servo bracket (Dog and Dino)
- 1 x Robot waist bracket – or make your own.

I obtained most of my parts for these robots on eBay. I did already have some servos but two of them were physically larger than the normal size of servos.

Here is the eBay ad for the robot kit that started everything.

ORDER DATE
Jun 18, 2015

ORDER TOTAL
US $74.92

See description

1 item sold by gadgetextreme

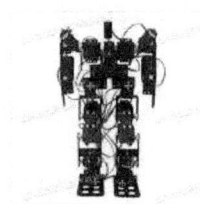

17DOF Biped Robot Educational Robot Kit Servo Bracket Ball Bearing Black e
(261684249440)

ITEM PRICE:
US $74.92

The robot kit did not come with any servos. To make the robot work I had to order 16 servo motors. With the two servos that I already had that brought my total to 18 servos. The two most common servos for making these robots are the Tower Pro MG995 and MG996. The MG996 is an improved version of the MG995.

Make sure you get the real "Tower Pro" servos and not some cheap imitations! The imitation servos sometimes will say "Towerd Pro" or their manufacturer name might be missing altogether.

ORDER DATE
Jul 17, 2015

ORDER TOTAL
US $21.18

+ US $1.00
shipping

1 item sold by czb6721960

4Pcs MG995 High Speed Digital Metal Gear 2BB Torque RC Servo
(130896119482)

ITEM PRICE:
US $21.18

For comparison purposes I also bought some MG996 and MG946R servos. They turned out to be some cheap imitations. One burned up on the first test run. Then I bought some MG958's. In my opinion they are the best servo for the price that you can get. The Tower Pro MG958 is a more powerful and a more expensive servo. However having at least two of the MG958's will really help with some of the heavier robots. The spider could not stand up without them.

ORDER DATE		ORDER TOTAL
Dec 03, 2015		US $51.60
		+ US $9.08 shipping
1 item sold by lagogla		

New T-Pro TPro MG958 Standard Digital Servo Metal Gear+ArmSet High Torque
(111724893554)
Quantity: 4

ITEM PRICE:
US $51.60

ⓘ Estimated delivery **Thu, Dec 10**
Tracking number: **9400111899223335690557**

In order to make the dinosaur with the long tail I ordered some more U shaped brackets. This is the ad for the longer U shaped bracket. The price is for two of them.

ORDER DATE		ORDER TOTAL
Oct 12, 2015		US $2.80
		Free shipping
1 item sold by better-seller01		

1Pcs Black Long U-Shaped Bracket Steering Gear Bracket Robot PTZ beus 2015
(181718199563)
Quantity: 2

ITEM PRICE:
US $2.80

ⓘ Estimated delivery **Wed, Oct 28 - Thu, Dec 03**
Tracking number: LK645419831CN

This next picture is of the shorter U shaped bracket. The price is for three.

ORDER DATE		ORDER TOTAL
Oct 13, 2015		US $4.11
		Free shipping
1 item sold by good4deal99		

1Pcs New Black Short U-Shaped Bracket Steering Gear Bracket Robot PTZ Good
(171504930007)
Quantity: 3

ITEM PRICE:
US $4.11

ⓘ Estimated delivery **Thu, Oct 29 - Fri, Dec 04**
Tracking number: **31172565424**

Here is the waist bracket. You will need two of them for the four legged robot and for the six legged robot. That is the case if you do not decide to make your own waist/chest bracket.

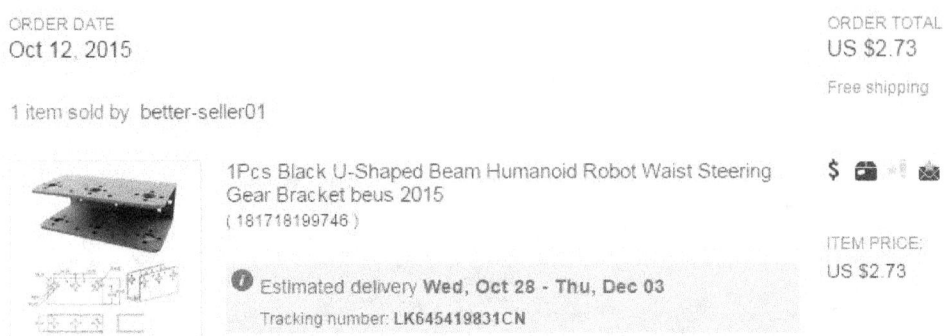

ORDER DATE
Oct 12, 2015

ORDER TOTAL
US $2.73

Free shipping

1 item sold by better-seller01

1Pcs Black U-Shaped Beam Humanoid Robot Waist Steering Gear Bracket beus 2015
(181718199746)

ITEM PRICE:
US $2.73

Estimated delivery **Wed, Oct 28 - Thu, Dec 03**
Tracking number: **LK645419831CN**

I also ordered some more servo brackets because some projects might require 18 servos and brackets. The price is for two of them.

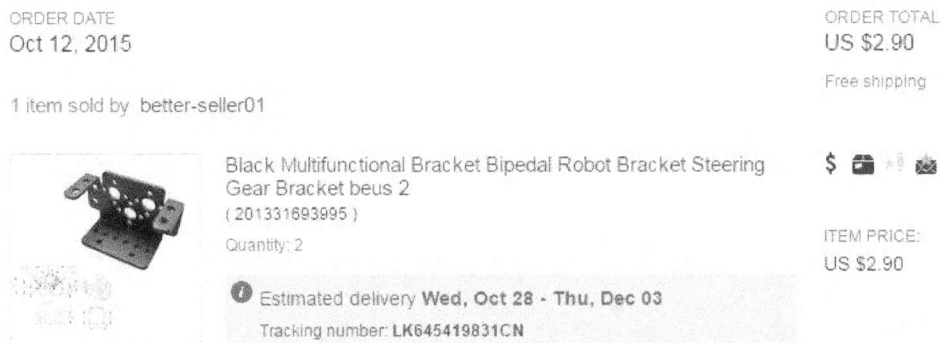

ORDER DATE
Oct 12, 2015

ORDER TOTAL
US $2.90

Free shipping

1 item sold by better-seller01

Black Multifunctional Bracket Bipedal Robot Bracket Steering Gear Bracket beus 2
(201331693995)
Quantity: 2

ITEM PRICE:
US $2.90

Estimated delivery **Wed, Oct 28 - Thu, Dec 03**
Tracking number: **LK645419831CN**

These are bearings that are needed for use with the normal single ended servos.

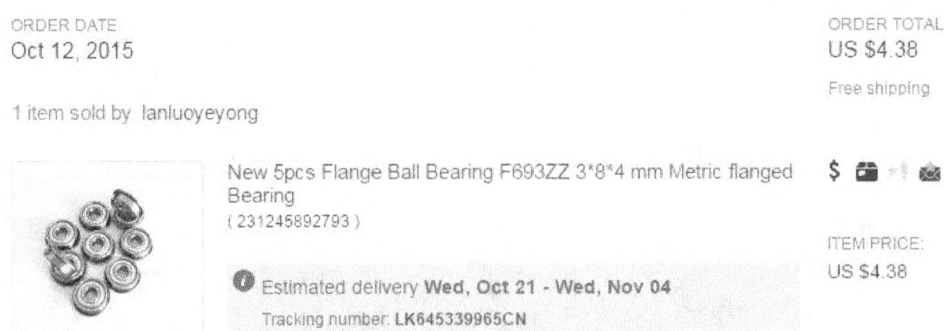

ORDER DATE
Oct 12, 2015

ORDER TOTAL
US $4.38

Free shipping

1 item sold by lanluoyeyong

New 5pcs Flange Ball Bearing F693ZZ 3*8*4 mm Metric flanged Bearing
(231245892793)

ITEM PRICE:
US $4.38

Estimated delivery **Wed, Oct 21 - Wed, Nov 04**
Tracking number: **LK645339965CN**

You might also need more servo "Horns". These have a lower profile and hence they work better with the larger MG958 servos.

ORDER DATE
Dec 03, 2015

ORDER TOTAL
US $7.47

+ US $2.25 shipping

1 item sold by upgradeindustries

Metal Servo Arm 25T Disc Metal Horns for MG995 MG996R RC Plane, Drone and Robots
(141599564024)
Quantity: 3

$ 📷 ⚡ ✉

ITEM PRICE:
US $7.47

ⓘ Estimated delivery Wed, Dec 09
Tracking number: 9400111699000702142640

Here is the sensor/servo shield to buy if you do not want to make your own servo shield.

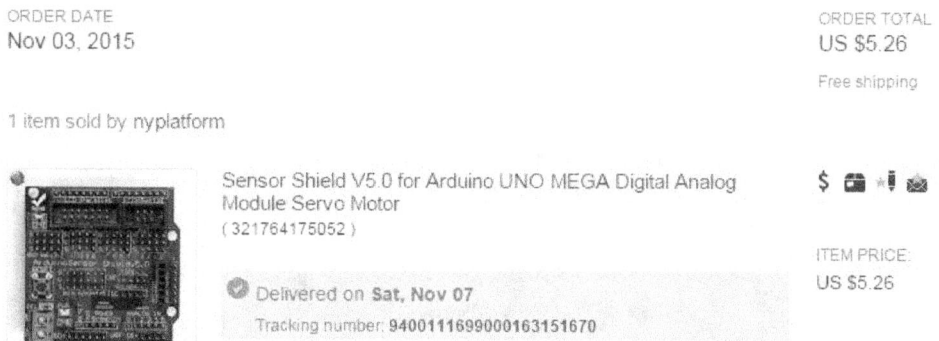

ORDER DATE
Nov 03, 2015

ORDER TOTAL
US $5.26

Free shipping

1 item sold by nyplatform

Sensor Shield V5.0 for Arduino UNO MEGA Digital Analog Module Servo Motor
(321764175052)

$ 📷 ⚡ ✉

ITEM PRICE:
US $5.26

✓ Delivered on Sat, Nov 07
Tracking number: 9400111699000163151670

Here is a wireless USB adapter that makes it easy to add remote control your robots. I made an adapter to connect the USB to TTL adapter and the transmitter adapter together. Then I glued them into the adapter.

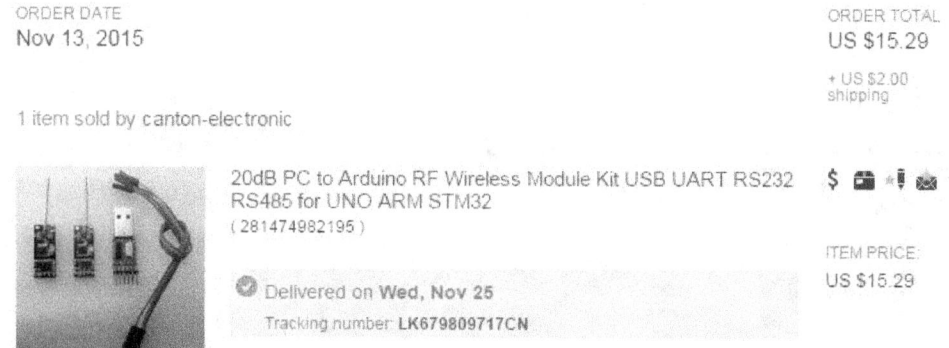

ORDER DATE
Nov 13, 2015

ORDER TOTAL
US $15.29

+ US $2.00 shipping

1 item sold by canton-electronic

20dB PC to Arduino RF Wireless Module Kit USB UART RS232 RS485 for UNO ARM STM32
(281474982195)

$ 📷 ⚡ ✉

ITEM PRICE:
US $15.29

✓ Delivered on Wed, Nov 25
Tracking number: LK679809717CN

Another thing that you might need for the larger robots are some servo extension cables like these.

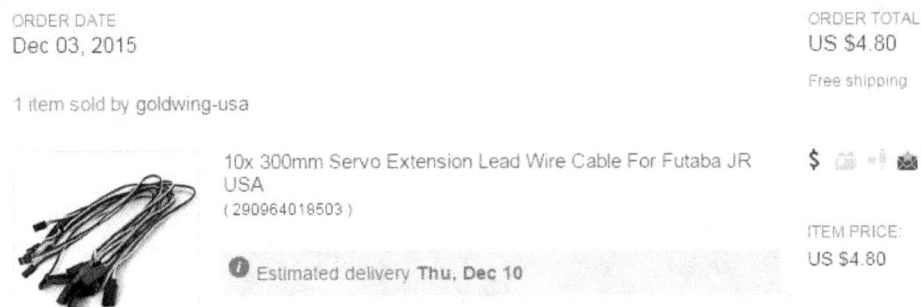

ORDER DATE
Dec 03, 2015

ORDER TOTAL
US $4.80

Free shipping

1 item sold by goldwing-usa

10x 300mm Servo Extension Lead Wire Cable For Futaba JR USA
(290964018503)

$

ITEM PRICE:
US $4.80

i Estimated delivery **Thu, Dec 10**

These are the optional metal hands. They give the 17 DOF humanoid robot the ability to pick things up.

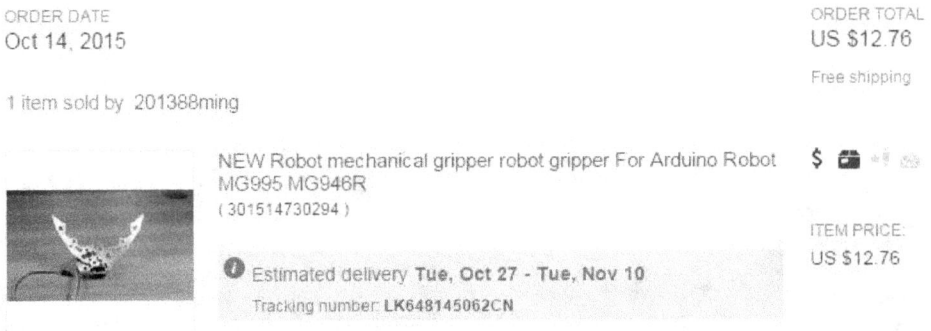

ORDER DATE
Oct 14, 2015

ORDER TOTAL
US $12.76

Free shipping

1 item sold by 201388ming

NEW Robot mechanical gripper robot gripper For Arduino Robot MG995 MG946R
(301514730294)

$

ITEM PRICE:
US $12.76

i Estimated delivery **Tue, Oct 27 - Tue, Nov 10**
Tracking number: **LK648145062CN**

If you prefer plastic hands these are an example of them. They can double for a mouth when making the dog or the dinosaur.

ORDER DATE
Sep 28, 2015

ORDER TOTAL
US $21.84

Free shipping

1 item sold by skt_flyer

Servo Grippers Robotic Crawling Robot Gripper Symmetry For Toy Car Robotic Model
(391242375172)

Quantity: 2

$

ITEM PRICE:
US $21.84

i Delivered on **Thu, Oct 15**
Tracking number: **LK628800563CN**

For some of the robot projects I made some of my own parts. How to make some of those parts will be covered under the robot that used them. However the mechanical design of my parts, and hence the programs in this book are completely compatible with the standard parts.

The typical waist/chest piece is made out of several brackets and is ugly in my opinion. I chose to make my own chest out of aluminum instead. Coming up is a picture of what the standard waist and chest arrangement looks like. That combination of brackets can be replaced by one bracket if you make it yourself. The home made chest bracket is completely interchangeable with the standard chest/waist assembly that is pictured below.

Another possible robot chest arrangement uses two of the waist brackets and an "L" shaped bracket for mounting the head. The head is a little lower in this design and may need some spacers to raise it up above the shoulders. Up next is a picture of that alternate waist and chest arrangement.

The home made chest should be made out of a three and a half inch long section of three inch wide by one inch tall "C" shaped aluminum channel. Holes are then drilled in it for mounting the legs, the head, and the shoulder servos. The hole spacing is roughly the same as the waist piece that is normally used in making these robots. In fact I have used the holes in the standard waist bracket to verify the locations of the homes in the home made chest.

This home made chest piece is roughly the same thing as you would get by bolting two of the normal waist bracket pieces together at their open ends. However this design will give you an empty "channel" where you can mount the Arduino Uno and the servo shield. Otherwise you have to mount the Arduino and shield to the back of the robot where it gets beat up every time the robot falls down.

Coming up next is a picture of the front and back sides of the home made chest piece. This picture was taken before they were spray painted black. You do not need to drill the middle three holes on the front side of the chest.

Coming up next is the mechanical drawing for the home made chest. This is actually an improved chest design that is more like the standard waist piece as far as the size and spacing of the holes is concerned. This drawing also shows the correct width of 3.5 inches to match the width of the standard waist bracket. I only had a 3 inch wide piece of aluminum on hand when I started making these robots so the chest piece in some of my pictures is a little narrower than it should have been.

The first chest piece that I made also had additional holes that were added to allow the mounting of the shoulder servo brackets using the larger screws. You might see that chest piece design in some of the pictures in this book. However that earlier design made the distance between the shoulders smaller than it should have been. As a result it is not a design that I would recommend making.

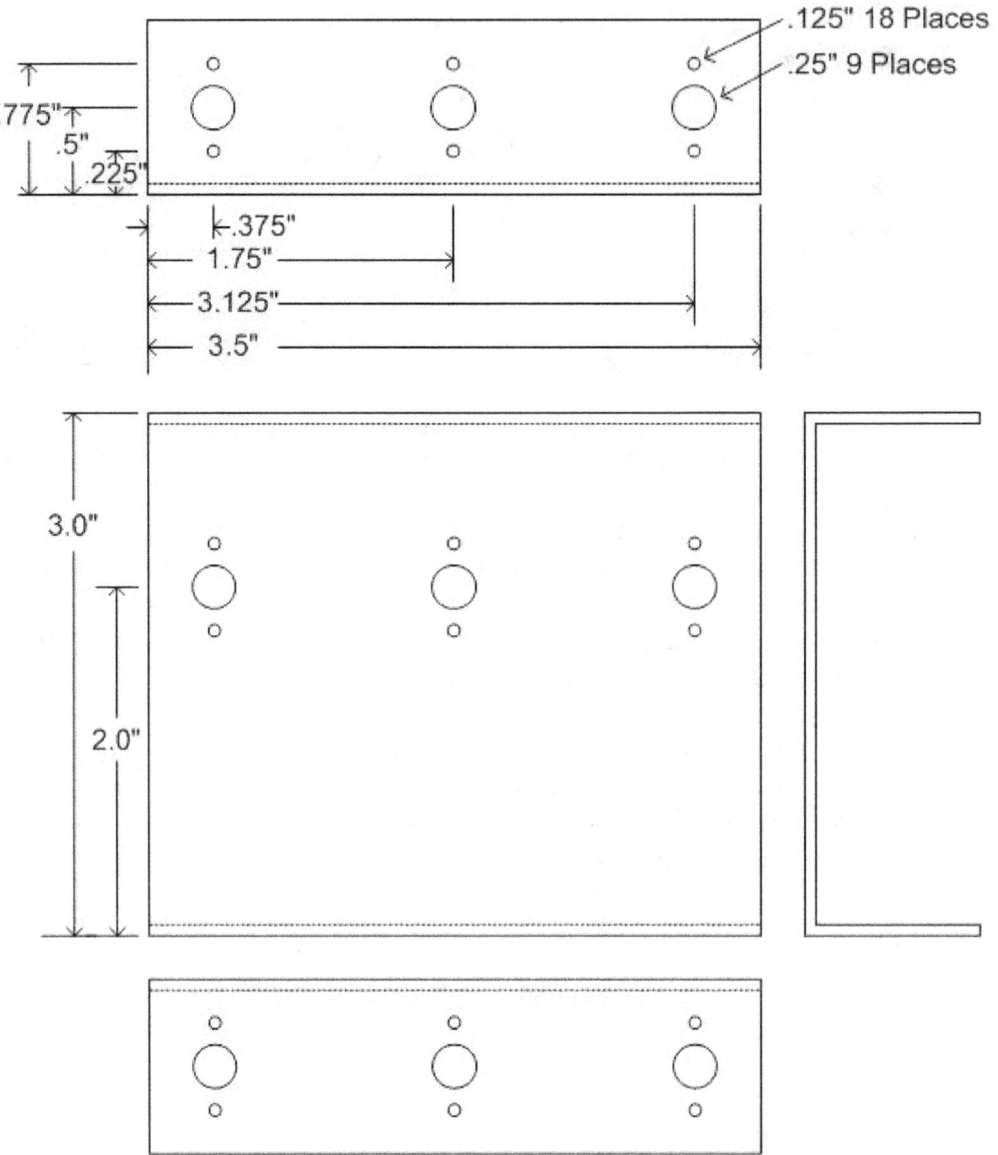

Another improvement that you can make for your robot is to give him a better looking head. I made a nice robot head out of a clear 3.2 inch (80mm) plastic Christmas tree globe. First you will need to layout the design on the globe with a marker. For the servo motor you will need an opening about 1.75 by .75 inches in size. Center it around the globes mounting hole.

I used a fine tooth coping saw to make the two longer cuts. Then wire cutters made the two shorter cuts to connect the two longer ones together.

Wire cutters are then used to trim the opening to where the globe will fit over the servo.

The finished cutout will also need a place on one side for the servo wires to fit through. This next picture shows the opening for the servo in the bulb. I actually had to make the opening a little larger to fit the circuit board through it.

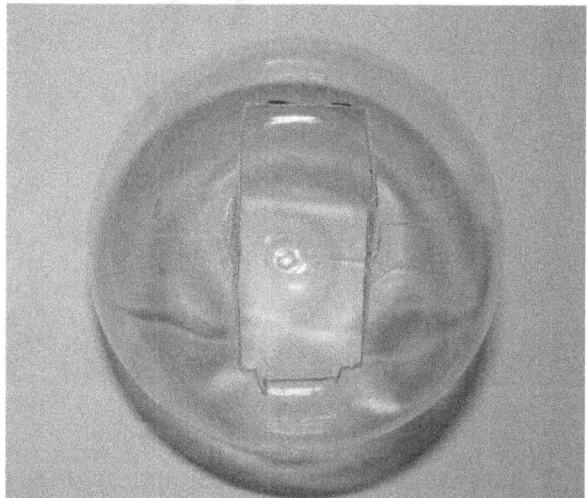

Next you will need to tape off where the opening will be for the robot's "face" to show through. Your circuit board will have to be filed to match the size and shape of this opening. Next you can lightly sand and spray paint the globe. Even the unpainted globe in this next picture looks better than the plain servo motor.

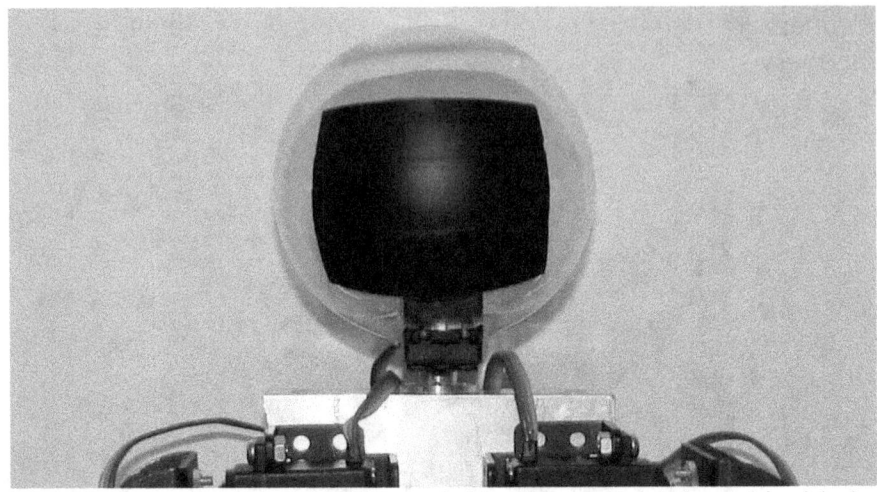

This next picture shows the painted globe with a circuit board inside it. LED's on the board form a mouth and eyes. The eyes are lit whenever power is on and the mouth lights when he receives commands.

The schematic of the face is really simple. I used the LED's in parallel. That requires that the LED's match each other. It they do not match you can have one LED that is much brighter or dimmer that the others.

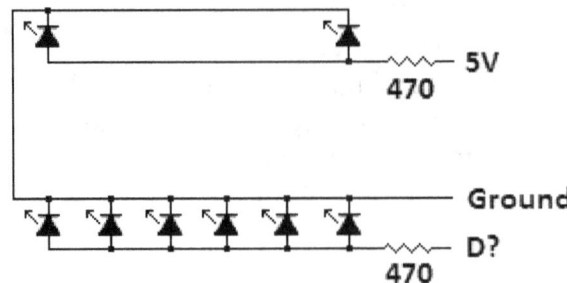

Chapter 3

Servo Controller Shield

You can purchase a generic servo controller shield for your Arduino but you can also easily make your own servo controller shield if you know how to solder. The servo shields construction is not that tricky as most of the servo connector pins are just soldered together in parallel rows.

The electrical design is fairly simple. Basically all of the pins furthest from the data pins go together to a ground connection binding post. Then all of the middle pins go together to an external power connector binding post (Usually 6 volts). Then all of the pins closer to the data pins go to the data pin that they are closest to.

There are a total of 14 data pins but the fist two data pins are usually not used. That is because D0 and D1 are used to communicate with the Arduino. If you connect a servo to them it will go through all kinds of contortions whenever you update the software in the Arduino.

Up next is a picture of the bottom view of the shield. Note in the picture that I connected the top two servo control lines back to D11 and D13. This arrangement allows for two servos to be connected to the same control pin. That will work for instances where you have two servos that are essentially doing the exact same thing. Using this trick you can then control more that the 12 servo limit.

Later on I added six more servo positions that are connected to the six analog input pins. If you buy a servo shield usually all 20 possible positions are already wired in the shield.

Here is the top view of the servo control shield.

In the last picture you can see an optional 100uF at 16 volt capacitor is connected from the power pins to the ground pins. This helps to reduce the amount of electrical noise on the power lines. It is still not safe to power the Arduino from the same power supply as the servos.

Here is a picture of a professional servo/sensor shield for comparison. The "GVS" stands for Ground Voltage and Signal. Note that it has jacks for LCD screens, Bluetooth, and several other things. The Bluetooth connector is located on the right side and it is offset. Bluetooth only uses the top four pins. They are 5 volts, Gnd, TX and RX. The unused two bottom pins are Gnd and 3 volts.

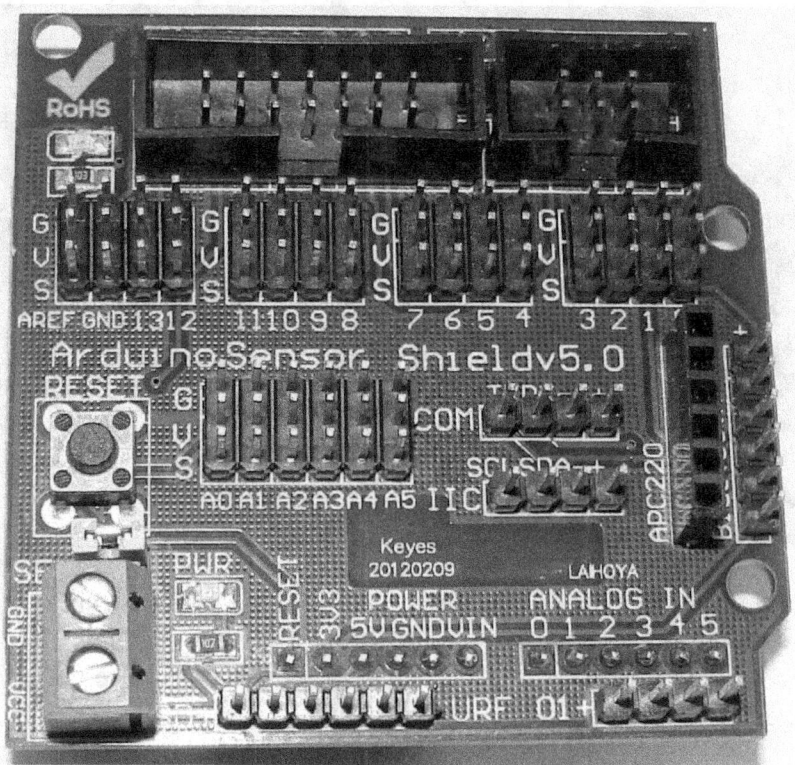

There is a design bug in the design of the servo/sensor shield. First you need to remove the green jumper to separate the 6 volts for the servo power at the screw terminals from the 5 volts logic for the Arduino. However even after you do this the servos connected to the analog (A0 to A5) servo pins are still connected to the 5 volt logic level! If those servos try to do any work at all they will overload the 5 volt regulator and shut the Arduino down.

The fix for the sensor shield is not that easy to do. You have to cut the 5 volt logic run in two places then jumper 5 volt power around the servo pins. Then you need to connect the servo pins back to the 6 volts from the screw terminals. These changes are shown in the next picture. Also there was a run that was burnt off the board in the top right corner of the picture that has been repaired.

Cut two runs here. Then add Jumpers for 5 Volt logic and for 6 Volt servo power.

Usually a nine volt battery is used to power the Arduino. A six volt battery pack consisting of four rechargeable AA batteries is used to power the servos. The four AA batteries actually deliver about five volts as rechargeable AA batteries are 1.2 volts each.

Coming up next is a picture of the batteries that you will need for wireless operation of your robots. At the top of the picture is a standard 9 volt battery with the needed adapter to power the Arduino. Below that are two different options for powering the servos.

Rechargeable batteries are preferred as you will drain a lot of batteries while you test out your programs. Also the rechargeable batteries put out nearly two amps of power for as much as an hour of operation. That would be 30 minutes of power at four amps of power usage.

For testing you can usually use a 6 volt 2 amp AC adapter. For my tests I used a 5 volt 10 amp power supply that I have turned up to about 5.6 volts. That way I did not have to keep recharging the batteries.

Chapter 4

Wireless Remote Control

Another type of wireless communications device to use for these projects is a 2.4 GHz radio frequency device called "Bluetooth". Probably the best wireless device to use for controlling things is "Bluetooth" because you can also control it with your telephone or anything else that can send data to a Bluetooth device. Bluetooth also emulates a serial port, so data can be sent and received just like it was a direct serial or USB connection.

When the Bluetooth receiver arrived in the mail it did not came with any real instructions. All that I had to work with was these specifications:

Default serial port setting : 9600 1
Pairing cod e: 1234
Running in slave role : Pair with BT dongle and master module
Coupled Mode : Two modules will establish communication automatically
        when they are powered up.
PC hosted mode : Pair the module with Bluetooth dongle directly as a
        virtual serial device.
Bluetooth protocol : Bluetooth Specification v2.0+EDR
Frequency : 2.4GHz ISM band
Modulation : GFSK(Gaussian Frequency Shift Keying)
Emission power : <=4dBm, Class 2
Sensitivity : <=-84dBm at 0.1% BER
Speed : Asynchronous: 2.1Mbps(Max) / 160 kbps,
Synchronous : 1Mbps/1Mbps
Security : Authentication and encryption
Profiles : Bluetooth serial port
CSR chip : Bluetooth v2.0
Wave band : 2.4GHz-2.8GHz, ISM Band
Protocol : Bluetooth V2.0
Power Class : (+6dbm)
Reception sensitivity: -85dBm

Voltage : 3.3 (2.7V-4.2V)
Current : Paring - 35mA, Connected - 8mA
Temperature : -40~ +105 Degrees Celsius
User defined Baud rate : 4800, 9600, 19200, 38400, 57600, 115200,
230400,460800,921600 ,1382400.
Dimension : 26.9mm*13mm*2.2mm

Setting up a Bluetooth wireless remote control is not an easy thing to do
the first time. It took me several hours to get it working properly. First,
start by plugging in the USB Bluetooth adapter to your PC and then
installing the needed drivers. Usually it will automatically install the
drivers over the Internet if you select that option and have a working
Internet connection.

Next you can power up the Bluetooth receiver module that will connect to
the Arduino. Only connect the power and ground pins to the receiver for
now. If you have the complete adapter with a voltage regulator built in it
will work on either 3.3 or 5 volts, otherwise it must have 3.3 volts to
operate. The receiver should start blinking a red LED to let you know that
it has power.

On the PC, go to "Control panel" and select the "Bluetooth devices" icon.
In the menu that comes up, select "add" and then check the box that says
"It is powered up and ready to be found" and then select "next". Then you
should see as the computer searching for Bluetooth devices.

Most of the Arduino interface Bluetooth devices will come up as "HC-
06". Select it, and at the prompt, enter the default pass code of "1234".
The "Add Bluetooth Device" screen then look something like the
following picture.

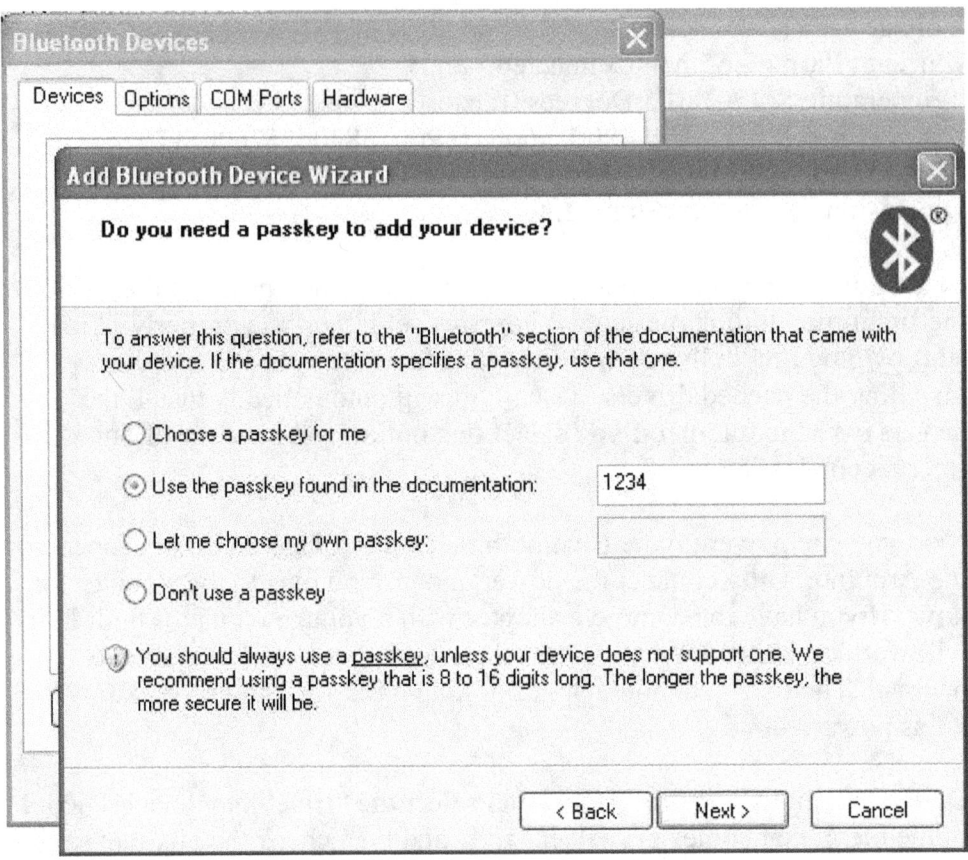

If the Bluetooth devices connect and everything works, the red light on the Bluetooth receiver will stop blinking and stay constantly lit. You should also get a message on the computer saying what serial port(s) have now been assigned to the Bluetooth device. It will look something like what you see in the next picture.

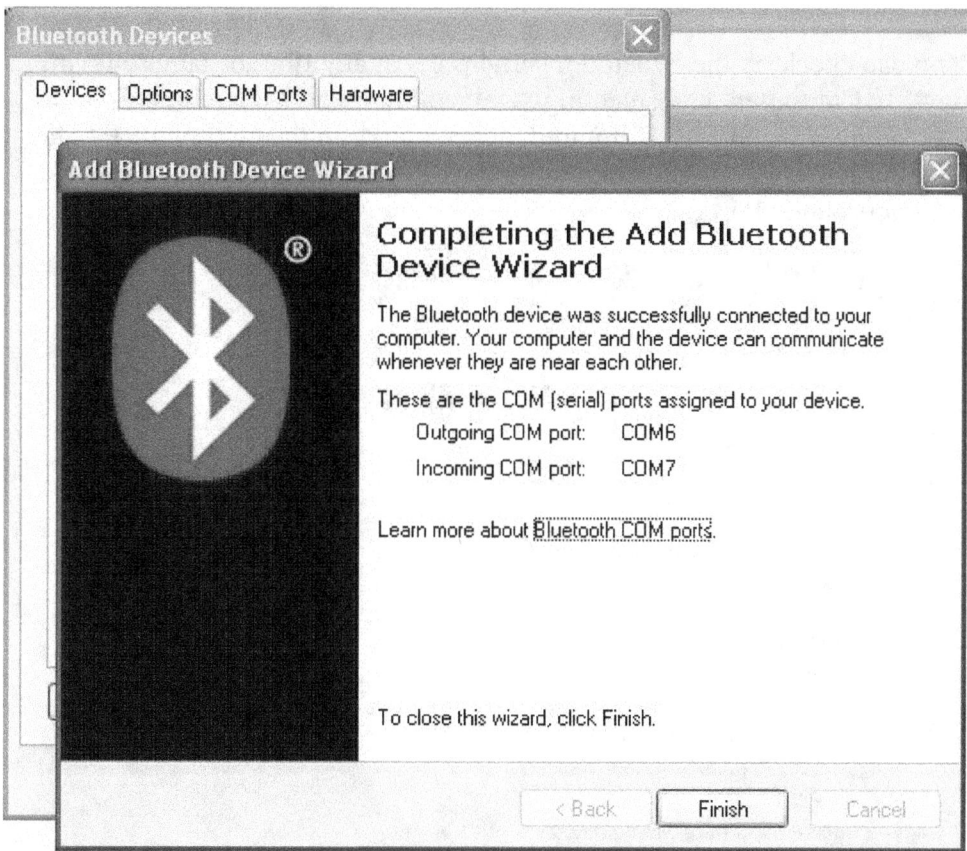

Next you need to load a sketch such as "Bluetooth test" into your Arduino. If you have a LCD connected, try the "Bluetooth to LCD" test program. Then connect the "TXD" pin of the Bluetooth receiver to the Arduino data pin D0. Note that this will block you from uploading any additional sketches until it is disconnected.

Next load a terminal program on the PC such as "Hyperterminal" and set it up for the first communications port that was assigned to your Bluetooth device, and "Serial," "9600 baud," "N," "8," and "1". Those settings should be the default settings. Save this instance of Hyperterminal as "Bluetooth" for future use.

At this point you should be able to type on your PC keyboard and see it on the LCD attached to the Arduino, or press "0" and "1" to see the LED on the Arduino pin 13 turn on and off. Congratulations, you have mastered connecting a PC to an Arduino via Bluetooth. I could not find any step by step instructions to do this anywhere!

You can check on the Bluetooth serial ports at any time by right clicking on "My Computer" and selecting "Device Manager". The screen should then show Bluetooth stuff in three locations like in the next picture. The Bluetooth adapter should come up under "Bluetooth Radios", "Network Adapters" and "Ports".

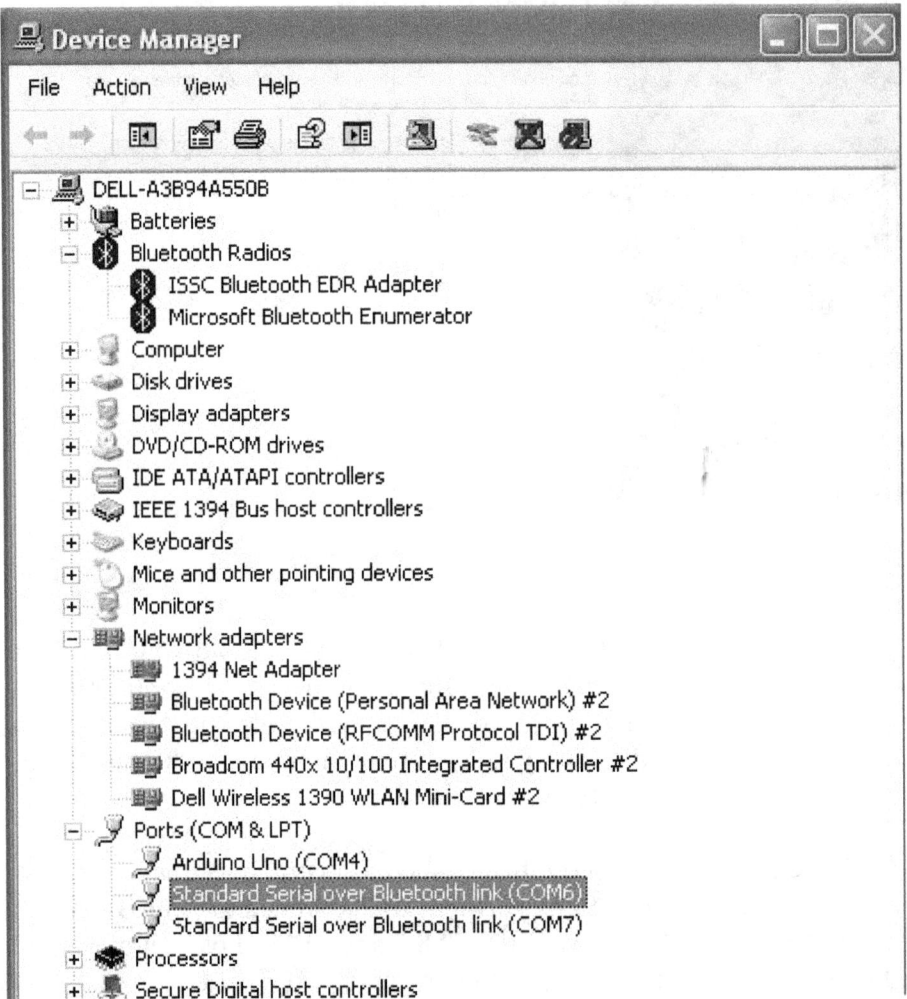

Up next is a picture showing how to connect the Bluetooth receiver to the Arduino. The Bluetooth module runs on 3.3 volts internally. **Do not connect D1 directly to the Arduino**, unless you have an adapter, use a 1K resistor in series, or simply do not use it. D1 is not needed to send commands from the PC to the Arduino. It is only needed to send responses back to the computer.

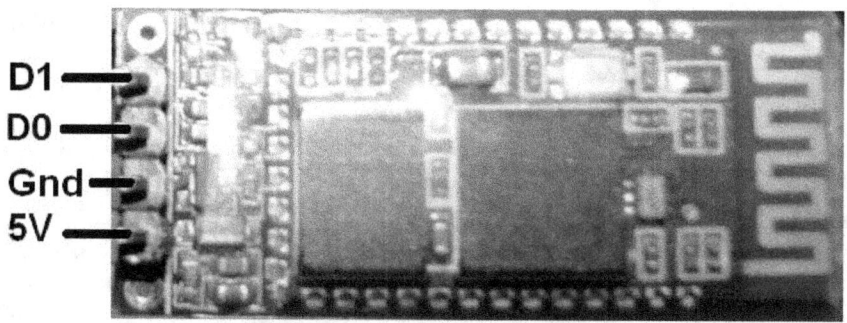

The keyboard commands can be very simple like this:
Forward = F
Back up = B
Right = R
Left = L
Stop = S
Clap = C

The programming for Bluetooth looks like this:

```
char INbyte;

void setup() {
  Serial.begin(9600);
}

void loop() {
  // read next available byte
  INbyte = Serial.read();
  // Execute the command
  if (INbyte == 'f'){
  // forward commands
  }
  if (INbyte == 'b'){
  // back commands
  }
  if (INbyte == 'r'){
  // right turn commands
  }
// Etc.
```

Here is a picture of the Bluetooth module attached to the sensor shield.

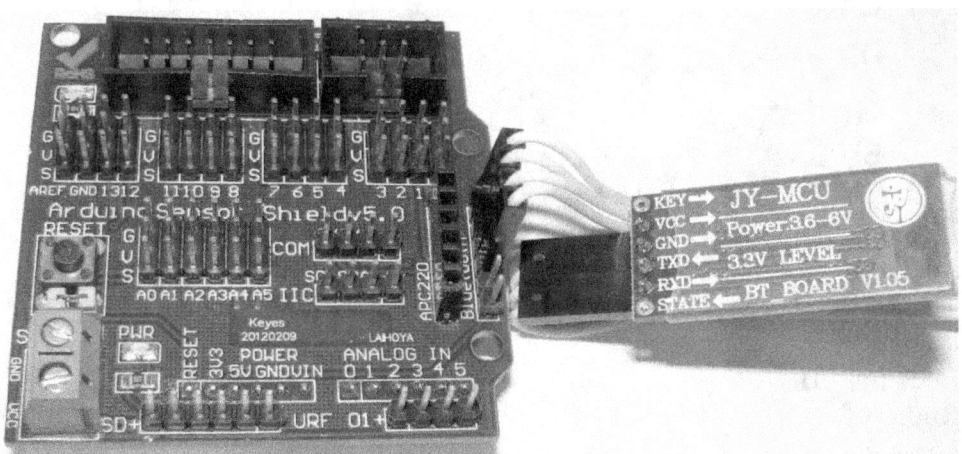

Please note that the Bluetooth connector is 6 pins but you only need to use four of them. The bottom two pins are not used. The top four bluetooth pins line up as VCC, GND, TXD and RXD.

Another option for wireless operation is to use wireless USB. This takes three parts, a USB to TTL adapter, and two bidirectional transceiver modules. I made my collection of parts into complete plug in adapters. This is a picture of the USB transmitter adapter.

To make the adapter I used a dual six pin header although you could glue two six pin headers together. Here is a picture showing how to wire the two adapters together. You might note that you are connecting TX to RX and vice versa. Once it is wired and working you might want to cover it with a removable glue to keep it together.

Next you will need to make an adapter for the receiver. For this adapter you will need two five pin headers.

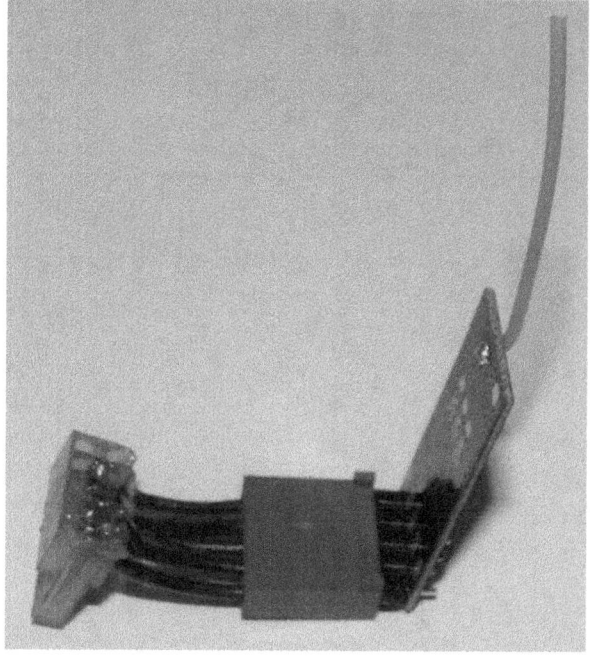

The wiring for this adapter is easier. The right two are direct and the next two - TX and RX must be crossed. The receiver adapter plugs into the Bluetooth jack on the servo/sensor shield once it is done.

Chapter 5

Walking Humanoid Robot

9 DOF

A simple walking robot only needs two servo motors for each leg. One servo acts like an ankle to tilts and lift the other leg off the ground. Then the second servo rotates to move the other free leg forward.

One on the easy things that you can make yourself for your robot is the feet. They can easily be made out of two 2.5 inches by five inches pieces of plastic or aluminum. The only necessary holes to be drilled are four holes for the servo bracket to be mounted to the feet. The holes should be beveled or countersunk on the bottom side so that flat head screws will be flush to the surface.

The servo mounting holes are 1.75 and 2.125 inches from the back edge of the foot. The holes are located 5/16 of an inches form the outside edges of the feet.

Up next is a picture of the cut and drilled feet. I used darker plastic to match up with the rest of the robot.

Here is a picture of the feet with the ankles attached to show how they will go together. The bracket from the bottom servo is a long U bracket.

The arms and legs are each assembled first without the shoulder servos. The arms and legs are then powered up and verified to be correct in their servo positions at 90 degrees. The shoulder servo and hip servo flanges are attached to the chest piece. Then the arms and legs are attached to the servos on the chest piece.

Coming up next is a picture of the assembled and working 9 DOF robot. The shoulder servos worked best if they are mounted in front of the chest piece. That arrangement leaves room for the Arduino and servo shield. It also makes it much easier to get to the leg servo mounting screws. I had to adjust them to tighten them a couple of times as they seem to work their way loose and then a leg falls off! The spacer in between the shoulder servos is just there as a decoration. It looked like something was needed so I put the spacer there.

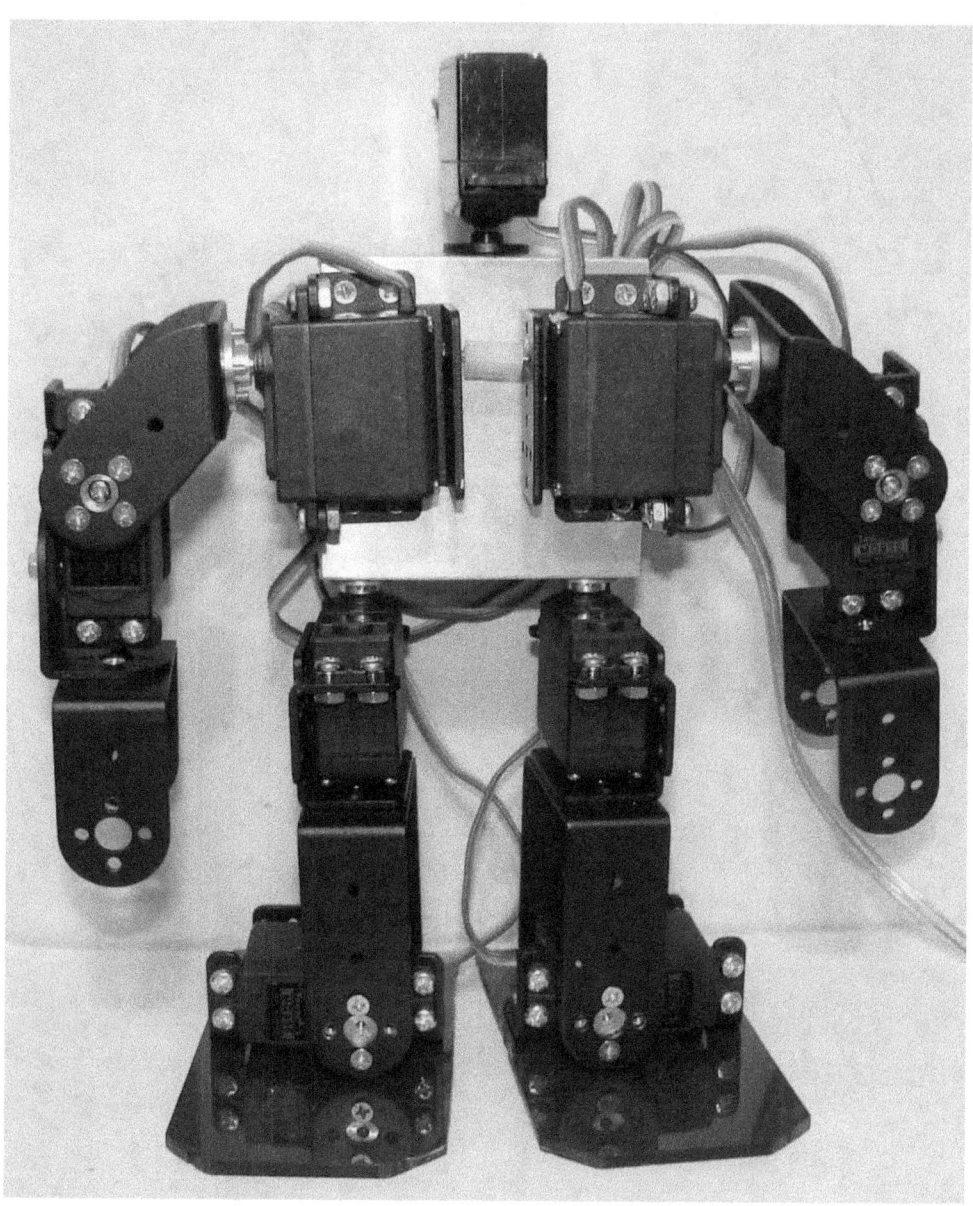

Up next is a picture of the back side of the 9 DOF robot showing the wiring. The servo wires are all a bit long for this smaller robot. They needed to be tied up somehow so that they do not get tangled into the moving parts. The Arduino fits crosswise. One of its mounting screws is also used to hold the shoulder servo bracket in place.

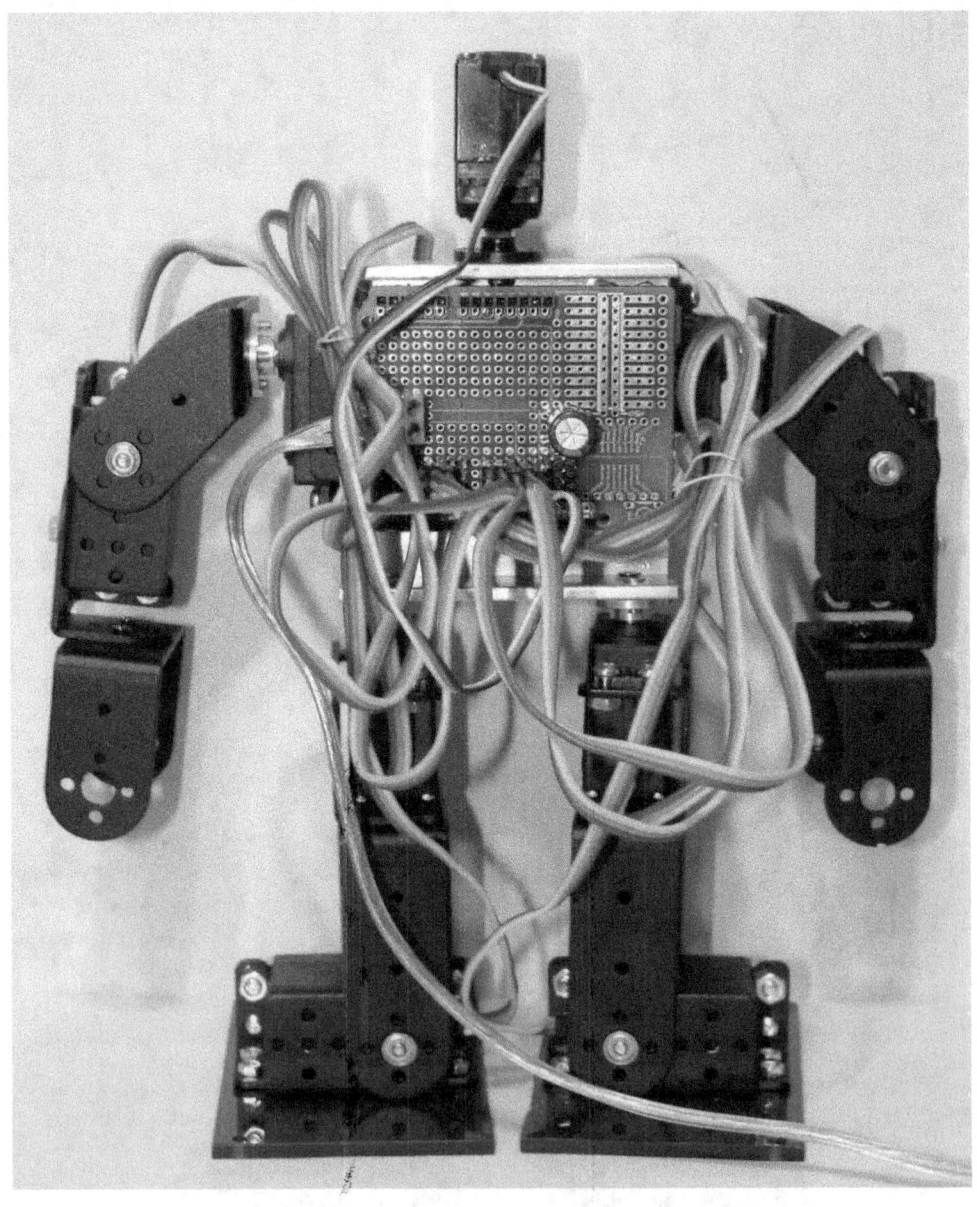

Here is the 9 DOF robot parts list:

- 8 x Multi-functional servo mounting bracket
- 2 x Short U-type servo bracket
- 2 x Long U-type servo bracket
- 2 x Angled U-type servo bracket
- 2 x L-type servo bracket
- 1 x Robot Waist bracket (not pictured, because I made my own)
- 2 x Foot Base
- 8 x Miniature Ball Radial Bearing
- 9 x Metal Servo Horn Flange
- Lots of Screws and nuts
- 9 x MG995 or MG996 Servo's.
- Arduino Uno and Servo Shield

This is the code to make him walk four steps forward and four steps backward. Then he will wave his arms and shake his head depending on what servos are attached.

```
// Humanoid 9 DOF demo
// Oct 2015 by Bob Davis
// Servos start at D2 as servo number 1
// Through D5 as servo number 4
// Then D8 as servo number 5
// Through D12 as servo number 9
// I skipped using D0, D1, D6 and D7

#include <Servo.h>
Servo servo1;  // Left Ankle
Servo servo2;  // Left Hip
Servo servo3;  // Right Ankle
Servo servo4;  // Right Hip
Servo servo5;  // Left arm
Servo servo6;  // Left Shoulder
Servo servo7;  // Right arm
Servo servo8;  // Right Shoulder
Servo servo9;  // Head
int pos = 0;   // variable to store the servo position

void setup() {
```

```
servo1.attach(2);  // Left Ankle - attaches the servo
servo2.attach(3);  // Left hip
servo3.attach(4);  // Right Ankle
servo4.attach(5);  // Right hip
servo5.attach(8);  // Left arm
servo6.attach(9);  // Left Shoulder
servo7.attach(10);  // Right arm
servo8.attach(11);  // Right Shoulder
servo9.attach(12);  // Head

servo1.write(90);  // Set zero positions
servo2.write(90);  //
servo3.write(90);  //
servo4.write(90);  //
servo5.write(90);  // Set zero positions
servo6.write(170);  //
servo7.write(90);  //
servo8.write(10);  //
servo9.write(90);  // Set zero positions
delay(200);
}

void StepForward() {
servo1.write(80);  //shift weight to left ankle
servo3.write(80);
delay(200);
servo2.write(110);  //left hip
servo4.write(110);  //right hip
delay(200);
servo1.write(90);  //level ankle
servo3.write(90);  //right ankle
delay(200);
servo1.write(100);  //Shift weight to right ankle
servo3.write(100);
delay(200);
servo2.write(70);  //left hip
servo4.write(70);  //right hip
delay(200);
servo1.write(90);  //level ankle
servo3.write(90);  //right ankle
delay(200);
```

```
}

void StepBackward(){
 servo1.write(80);  //shift weight to left ankle
 servo3.write(80);
 delay(200);
 servo2.write(70);  //left hip
 servo4.write(70);  //right hip
 delay(200);
 servo1.write(90);  //level left ankle
 servo3.write(90);  //level right ankle
 delay(200);
 servo1.write(100);  //Shift weight to right ankle
 servo3.write(100);
 delay(200);
 servo2.write(110);  //left hip
 servo4.write(110);  //right hip
 delay(200);
 servo1.write(90);  //level ankle
 servo3.write(90);  //right ankle
 delay(200);
}

void WaveArms(){
 servo6.write(10);
 delay (500);
 servo8.write(170);
 delay (500);
 servo5.write(50);
 delay (200);
 servo5.write(130);
 delay (200);
 servo7.write(130);
 delay (200);
 servo7.write(50);
 delay (200);
 servo5.write(50);
 delay (200);
 servo5.write(130);
 delay (200);
 servo7.write(130);
```

```
  delay (200);
  servo7.write(50);
  delay (200);
  servo9.write(40);  // head
  delay (500);
  servo9.write(140);
  delay (500);
  servo5.write(90);
  servo6.write(170);
  servo7.write(90);
  servo8.write(10);
  servo9.write(90);
}

//Main Control loop
void loop()  {
  StepForward();
  StepForward();
  StepForward();
  StepForward();
  servo2.write(90);  //left hip
  servo4.write(90);  //right hip
  delay(500);
  StepBackward();
  StepBackward();
  StepBackward();
  StepBackward();
  servo2.write(90);  //left hip
  servo4.write(90);  //right hip
  delay(500);
  WaveArms();
  delay(500);
}
// end of program
```

Chapter 6

Walking Humanoid Robot

13 DOF

The ultimate project is to make a human like walking robot. This next project is a walking "Humanoid" robot that walks more like the way we walk than the first robot did. This project will require a total of 13 servo motors. We will be adding a right and left "knee" and "hand" servos to the previous design.

Here is the 13 DOF robot parts list:

- 12 x Multi-functional servo mounting brackets
- 6 x Short U-type servo bracket
- 2 x Long U-type servo bracket
- 2 x Angled U-type servo bracket
- 2 x Flat servo bracket (Hands)
- 4 x L-type servo bracket
- 1 x Robot Waist bracket and/or Chest bracket
- 2 x Foot Base
- 12 x Miniature Ball Radial Bearings
- 13 x Metal Servo Horn Flanges
- Lots of screws and nuts
- 13 x MG995 or MG996 Servo's.
- Arduino Uno and Servo Shield

You can start building your robot by assembling the feet. Here is a picture showing how to assemble the feet. Use the flat head larger M4 sized screws as can be seen on the right side of this picture. Mount the bracket to the feet then mount the servo to the bracket. The left foot in the picture is showing the bottom side of the feet. The right foot shows the top side of the feet.

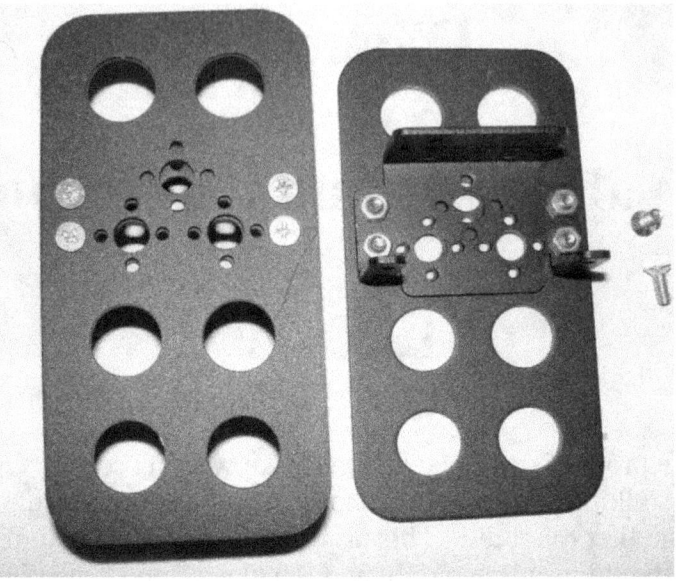

Here is what the feet look like once they are assembled. The left foot shows the back side of the feet and the right foot is showing the front view of the foot. This picture actually shows the wrong size of screws being used to mount the servos to their brackets.

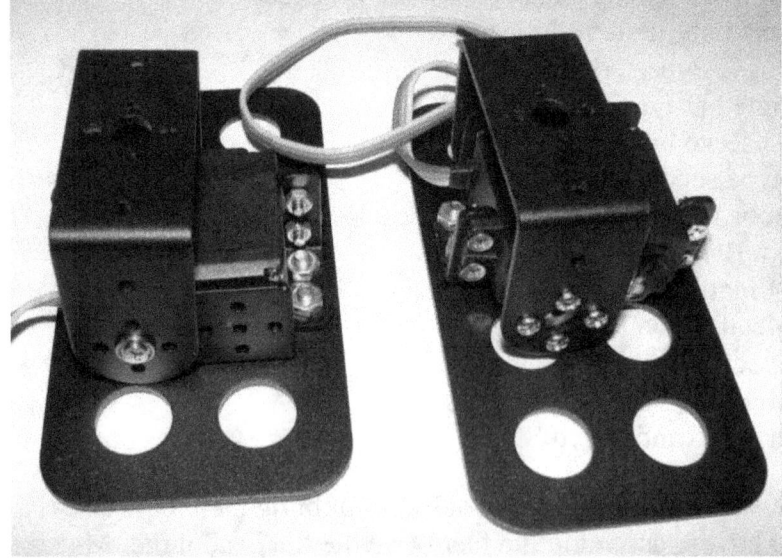

There are two ways to mount the bearing as can be seen on the left in the above picture. You can put the screw and bearing in first and then bend the "U" metal bracket over the screw and bearing. The, not so easy, alternate method is to use tweezers to hold the nut in place and then put the screw in as seen above. I did that a couple of times then switched to putting the screw in before installing the servo.

When I started assembling the legs I used the wrong screws to hold the servo's in place. You are supposed to use the M4 screws (A little bigger than a 6-32 screw) but I used the M3 screws (about the size of a 4-40 screw) instead. I had to change all of the servo mounting screws.

Also my guess is that the four screws used to hold the servo flange to the "U" shaped bracket are supposed to be flat head screws. That is based on the quantity of screws that were provided with the kit. The rounded or pan head screws are for the center of the flange to fasten it to the servo motor. The robot would look better if there were enough of the rounded head screws for fastening all of the flanges in place. The only place that really needs a flat head screw is for holding the bearing in place.

For this robot I used the same chest piece as was used for the 9 DOF humanoid robot. I did improve it a little bit with better hole placements for the hips. For assembly both the shoulder and hip servos are mounted to the chest. Then the arms and legs are assembles and tested. Then the arms and legs are attached to the shoulder and hip servos that are already attached to the chest.

The new leg assemblies have knee joints added in the middle. Also the hip joint now can move forward and backwards instead of moving in a circular motion. I used one long and one short "U" shaped bracket for connecting from the knee to the hip. Some other designs use two short "U" brackets and some designs use two long "U" brackets. I used one of each length to try to ensure compatibility with both of the other designs.

Coming up next is a picture of the assembled and working 13 DOF humanoid robot. Note the use of the home made chest bracket once again. This time he has a waist bracket and the chest bracket. That makes him about one inch taller. The waist bracket also provides a place where you can put his batteries. Also this time he has a "head" on his shoulders.

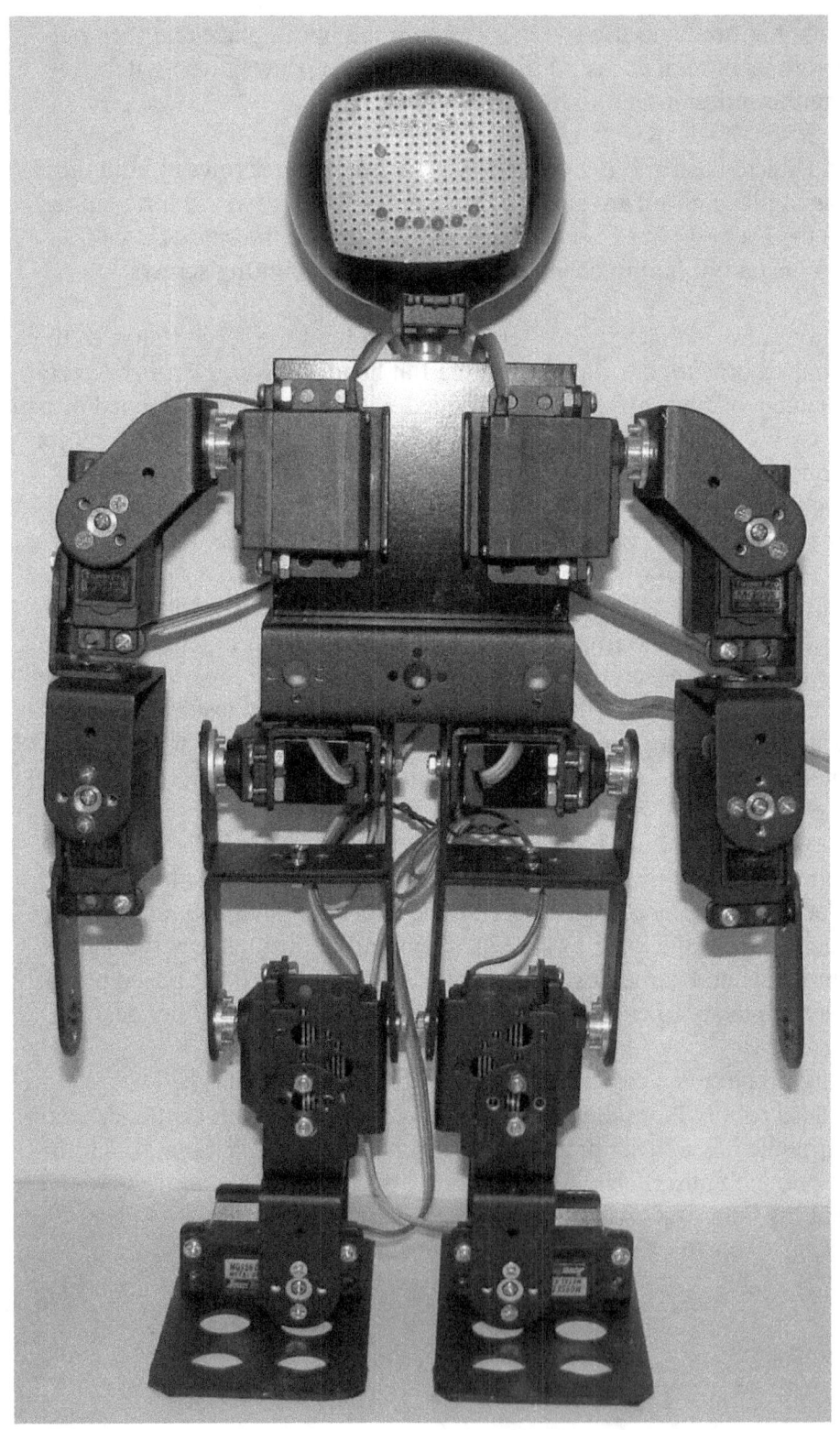

This next picture shows the back view of the 13 DOF robot with the servo numbers labeled to show you how to wire them up. Some sort of wire ties are needed to hold the wires in place.

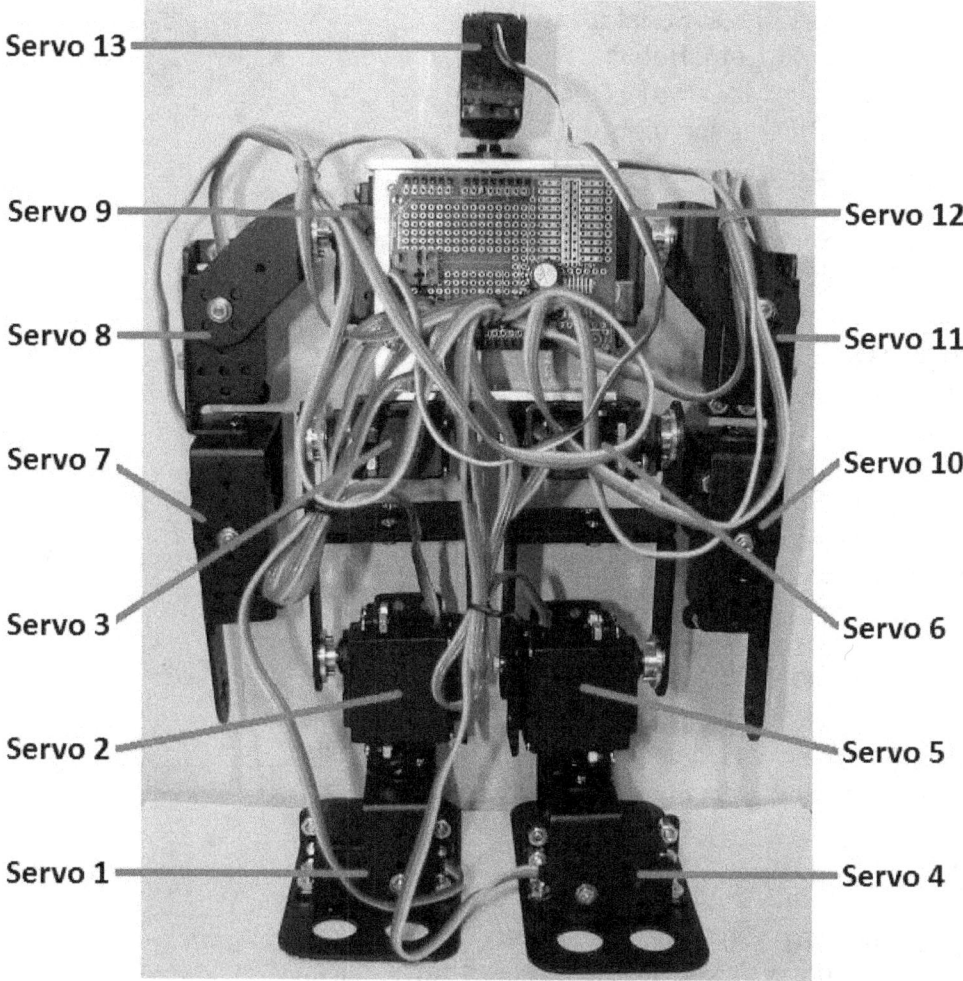

How do you get past the 12 servo limit? Notice that the head servo is connected to A1. I modified "Servo.h" to allow 13 servos per timer as seen in an earlier chapter.

For the robot moving forward and backwards is normal. Turning is a little more difficult. What he does to turn is to move a foot forward then drag it back into place. Depending upon the amount of drag created he may turn a little bit or a lot. Turning is not very accurate.

Here is the code to make him walk. Note that I am using arrays for walking forward, backwards and for turning. Using servo arrays makes complex movements easier to adjust and results in smoother movements. That is because there are more steps that allow for finer resolution.

```
// Humanoid 13 DOF Serial using arrays
// September 2015 by Bob Davis
// Mapped as D2 is servo1 is the "1" key
// Mapped as D3 is servo2 is the "2" key
// Etc.

#include <Servo.h>
// Define our Servo's
Servo servo1; // Left Ankle
Servo servo2; // Left Knee
Servo servo3; // Left Hip
Servo servo4; // Right Ankle
Servo servo5; // Right Knee
Servo servo6; // Right Hip
Servo servo7; // Left Hand
Servo servo8; // Left Elbow
Servo servo9; // Left Shoulder
Servo servo10; // Right Hand
Servo servo11; // Right Elbow
Servo servo12; // Right Shoulder
Servo servo13; // Head
int twait=100; // Sets time delay between steps

int walkf[6][10] = {
// 0   1    2    3    4    5    6   7   8   9
{ 95, 100, 110, 100, 100, 95, 90, 90, 85, 85}, // Left Ankle
{ 90, 90,  90,  85,  80,  75, 70, 70, 75, 80}, // Knee
{ 90, 90,  90,  85,  80,  75, 70, 70, 75, 80}, // Hip
{ 95, 100, 100, 100, 100, 95, 90, 90, 85, 85}, // Right Ankle
{ 90, 90,  90,  85,  80,  75, 70, 70, 75, 80}, // Knee
{ 90, 90,  90,  85,  80,  75, 70, 70, 75, 80}, // Hip
};

int walkb[6][10] = {
// 0   1    2    3    4    5    6   7   8   9
{ 95, 100, 110, 110, 100, 95, 90, 90, 85, 85}, // Left Ankle
```

```
{ 90,  90,  90,  95, 100, 105, 110, 110, 105, 100}, // Knee
{ 90,  90,  90,  95, 100, 105, 110, 110, 105, 100}, // Hip
{ 95, 100, 100, 100, 100,  95,  90,  90,  85,  85}, // Right Ankle
{ 90,  90,  90,  95, 100, 105, 110, 110, 105, 100}, // Knee
{ 90,  90,  90,  95, 100, 105, 110, 110, 105, 100}, // Hip
};

int turnr[6][10] = {
// 0   1   2   3   4   5   6   7   8   9
{ 95, 105, 110, 100, 100,  95,  90,  90,  90,  90}, // Left Ankle
{ 90,  90,  90,  85,  80,  75,  70,  70,  75,  80}, // Knee
{ 90,  90,  90,  85,  80,  75,  70,  70,  75,  80}, // Hip
{ 90,  95, 100, 100, 100,  95,  90,  90,  90,  90}, // Right Ankle
{ 90,  90,  90,  85,  80,  75,  70,  75,  80,  85}, // Knee
{ 90,  90,  90,  85,  80,  75,  70,  75,  80,  85}, // Hip
};

int turnl[6][10] = {
// 0   1   2   3   4   5   6   7   8   9
{ 85,  75,  70,  80,  80,  85,  90,  90,  90,  90}, // Left Ankle
{ 90,  90,  90,  95, 100, 105, 110, 110, 105, 100}, // Knee
{ 90,  90,  90,  95, 100, 105, 110, 110, 105, 100}, // Hip
{ 85,  75,  70,  80,  80,  85,  90,  90,  90,  90}, // Right Ankle
{ 90,  90,  90,  95, 100, 105, 110, 105, 100,  95}, // Knee
{ 90,  90,  90,  95, 100, 105, 110, 105, 100,  95}, // Hip
};

void setup(){
  Serial.begin(9600);
  Serial.print("Ready");
  servo1.attach(2); // servo on digital pin 2
  servo2.attach(3); // servo on digital pin 3
  servo3.attach(4); // servo on digital pin 4
  servo4.attach(5); // servo on digital pin 5
  servo5.attach(6); // servo on digital pin 6
  servo6.attach(7); // servo on digital pin 7
  servo7.attach(8); // servo on digital pin 8
  servo8.attach(9); // servo on digital pin 9
  servo9.attach(10); // servo on digital pin 10
  servo10.attach(11); // servo on digital pin 11
  servo11.attach(12); // servo on digital pin 12
```

```
    servo12.attach(13); // servo on digital pin 13
    servo13.attach(14); // servo on digital pin 14 (A0)
}

void loop(){
  if(Serial.available()) {
    char sinch=Serial.read();
    // Test out the servos
    if (sinch=='1') {
      servo1.write(50); //1 Key
      servo4.write(50);
      delay(twait*2);
      servo4.write(130);
      servo1.write(130);
      delay(twait*2);
    }
    if (sinch=='2') {
      servo2.write(50); //2 Key
      servo5.write(50);
      delay(twait*2);
      servo5.write(130);
      servo2.write(130);
      delay(twait*2);
    }
    if (sinch=='3') {
      servo3.write(50); //3 Key
      servo6.write(50);
      delay(twait*2);
      servo6.write(130);
      servo3.write(130);
      delay(twait*2);
    }
    if (sinch=='4') {
      servo7.write(50); //4 Key
      servo10.write(50);
      delay(twait*2);
      servo10.write(130);
      servo7.write(130);
      delay(twait*2);
    }
    if (sinch=='5') {
```

```
    servo8.write(50); //5 Key
    servo11.write(50);
    delay(twait*2);
    servo11.write(130);
    servo8.write(130);
    delay(twait*2);
  }
  if (sinch=='6') {
    servo9.write(50); //6 Key
    servo12.write(50);
    delay(twait*2);
    servo12.write(130);
    servo9.write(130);
    delay(twait*2);
  }
  if (sinch=='7') {
    servo13.write(50); //7 Key
    delay(twait*2);
    servo13.write(130);
    delay(twait*2);
  }
  // Sequential operations
  if (sinch=='c') {
    servo7.write(150); // c key=Clap hands
    servo8.write(150);
    servo9.write(30);
    servo10.write(30);
    servo11.write(30);
    servo12.write(150);
    servo13.write(50); // head
    delay(twait*2);
    servo13.write(130);
    delay(twait*2);
  }
  if (sinch=='r') { // right key;
    for (int i=0; i<10; i++){
      servo1.write(turnr[0][i]);
      servo2.write(turnr[1][i]);
      servo3.write(turnr[2][i]);
      servo4.write(turnr[3][i]);
      servo5.write(turnr[4][i]);
```

```
      servo6.write(turnr[5][i]);
      delay(twait);
    }
  }
  if (sinch=='b') {  // back key
    for (int i=0; i<10; i++){
      servo1.write(walkb[0][i]);
      servo2.write(walkb[1][i]);
      servo3.write(walkb[2][i]);
      servo4.write(walkb[3][i]);
      servo5.write(walkb[4][i]);
      servo6.write(walkb[5][i]);
      delay(twait);
    }
  }
  if (sinch=='l'){ // left key;
    for (int i=0; i<10; i++){
      servo1.write(turnl[0][i]);
      servo2.write(turnl[1][i]);
      servo3.write(turnl[2][i]);
      servo4.write(turnl[3][i]);
      servo5.write(turnl[4][i]);
      servo6.write(turnl[5][i]);
      delay(twait);
    }
  }
  if (sinch=='f') {  // forward key
    for (int i=0; i<10; i++){
      servo1.write(walkf[0][i]);
      servo2.write(walkf[1][i]);
      servo3.write(walkf[2][i]);
      servo4.write(walkf[3][i]);
      servo5.write(walkf[4][i]);
      servo6.write(walkf[5][i]);
      delay(twait);
    }
  }
  servo1.write(90); // Return to zero positions
  servo2.write(90);
  servo3.write(90);
  servo4.write(90);
```

```
      servo5.write(90);
      servo6.write(90);
      servo7.write(90);
      servo8.write(90);
      servo9.write(90); // Left Shoulder down
      servo10.write(90);
      servo11.write(90);
      servo12.write(90); // Right Shoulder down
      servo13.write(90);
  }
} // End of program
```

Chapter 7

Walking Humanoid Robot

17 DOF

Once again we will be using the home made breastplate but it is functionally the same as the one that is made out of several parts. This robot features four servos per leg. I think that four servos per leg is all that is needed. Some designs have five servos per leg. That limited number of servos when combined with the arms exceeds the limits of using the 12 Arduino digital pins for controlling the servos.

New for this project are some "real" hands. There are hands available in metal and plastic. There are hands available that take a dual ended servo and ones that take a normal servo. There are a lot of options to choose form. I decided to try both a plastic and a metal hand that take normal servos. Then I also will show a possible hand made with ordinary brackets.

The plastic hands require the use of a mini servo. However they included the servo with the kit. The plastic parts came covered in brown paper that had to be removed. The instructions were entirely in Chinese. You can use the numbers in the instructions to guess what size of screw and what size of spacer goes where.

Coming up is a picture of the various parts that came in the kit to make the hands. In the plastic bag of screws there are two different lengths of spacers. According to the instructions the longer spacers go towards the top of the hands.

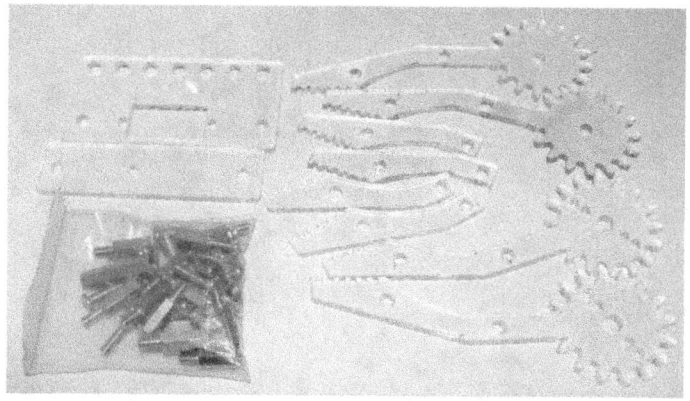

This is what an assembled plastic hand looks like.

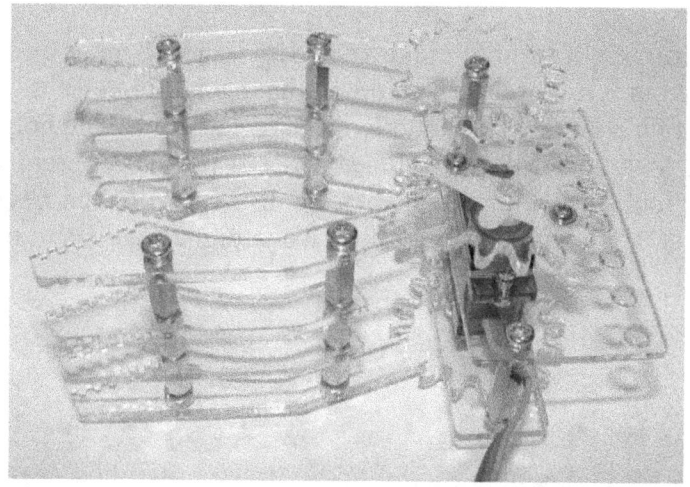

Attaching the plastic hands to the arms turned out to be a little tricky. I had to file the holes in the new hand brackets to slightly increase their spacing to match the spacing of the holes in the servo mounting bracket. Also, since the hands are identical, in order to make them "right" and "left" I had to use different mounting holes. One hand was mounted using the two end holes and the other hand used the third and fourth holes. It was also a little tricky to get the hands gears to clear the perpendicular portion of the servo mounting holes.

The metal hands did not come with a servo. However they fit a standard sized servo. The alignment is a bit rough as the height of the servo does not line up perfectly with the adjacent gear. Also the gears mesh together a little bit too tight. Here is a picture showing the metal hands with a front and back view.

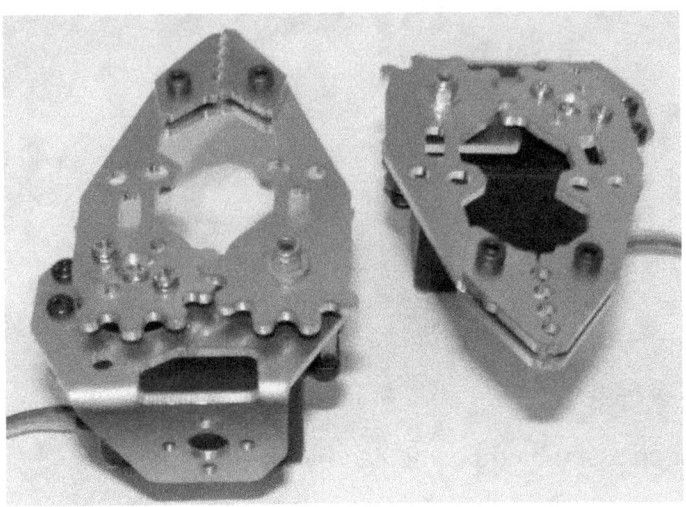

Last of all you might want to make your own hand. It takes a few brackets to make one but it might do if you do not have anything else to use. Here is a picture of a home made hand next to a metal hand. The home made hand would look neater if one of the straight brackets were replaced with a bracket that looks like four fingers. Basically you will need a servo, a servo bracket, a 90 degree bracket and two straight brackets to make a simple hand.

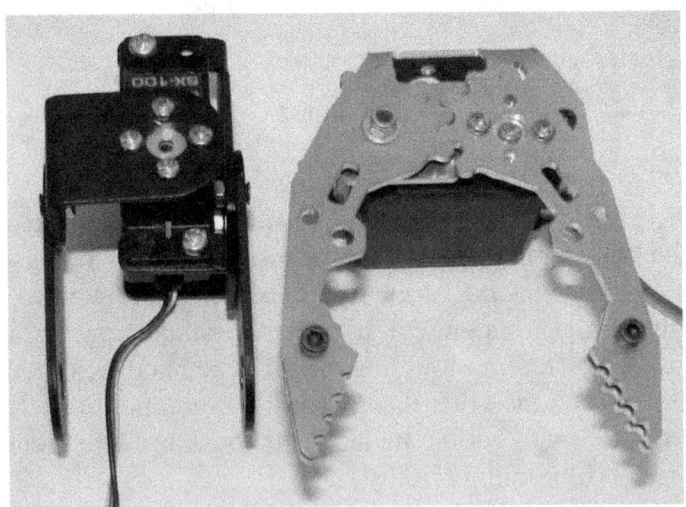

For this robot we are also adding another servo in the legs. It gives the hips a right-left movement as well as the already existing front to back hip movement. I also used two of the long U brackets for the middle part of his legs. I was trying to make him as tall as possible so that his new hands would not drag on the ground when he walks! Here is a picture of the new legs showing both a back view and a front view. You might notice the stripes on the new MG958 servo's.

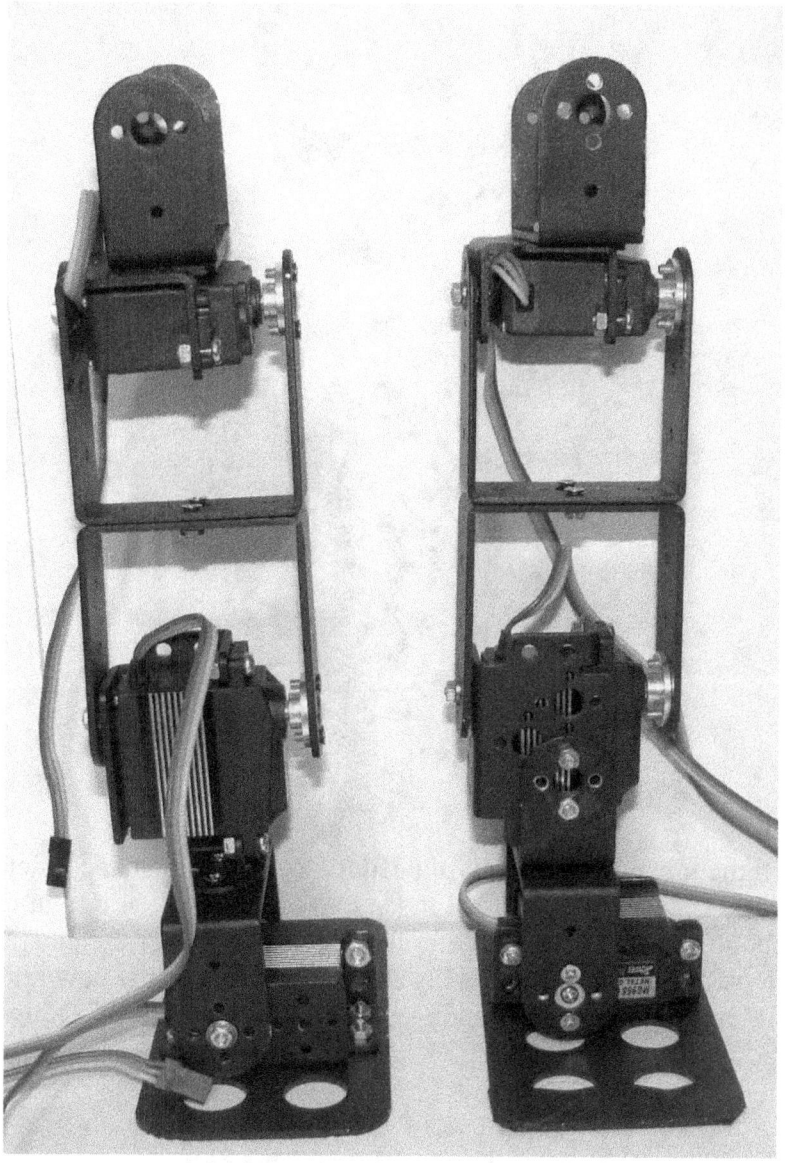

I also changed the chest design for this robot. I added another waist piece to it to make it about an inch taller. It is still functionally the same as the multi bracket chest but with a second waist piece. You could get similar results by mounting the shoulder servo brackets a little higher on the angle brackets and by adding some one inch spacers to make the head a little higher.

To attach the metal hands you will need to add two "L" shaped metal angle brackets. I had run out so I made two out of some aluminum angle pieces about one inch long. I prefer the metal hands over the plastic ones as they look a little "tougher". This next picture shows the new arms with his metal hands attached and ready to be used to make the 17 DOF humanoid robot.

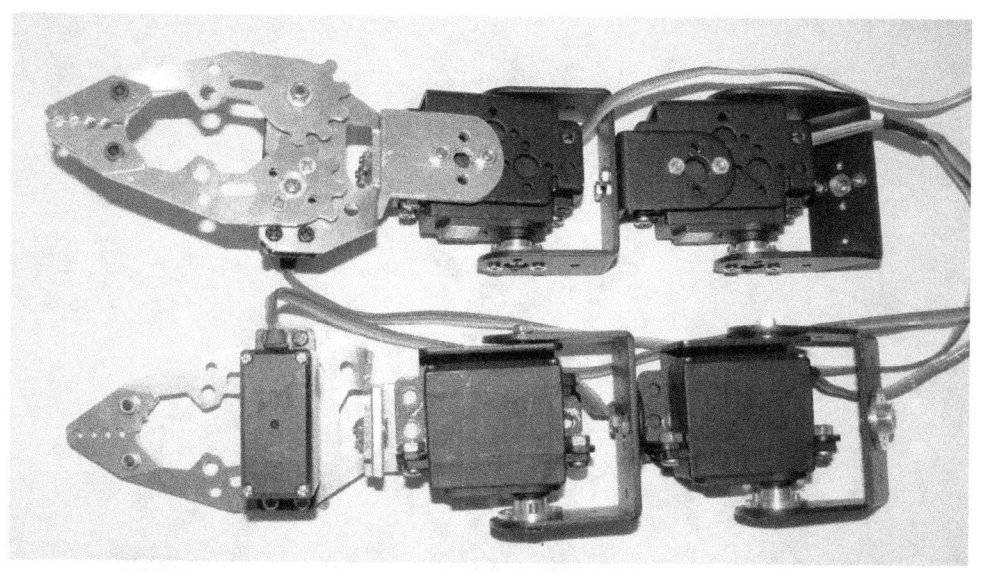

Here is the 17 DOF robot parts list:
- 16 x Multi-functional servo mounting bracket
- 6 x Short U-type servo bracket
- 4 x Long U-type servo bracket
- 2 x Angled U-type servo bracket
- 0 x Flat servo bracket
- 6 x L-type servo bracket
- 1 x Robot waist bracket
- 1 x Robot chest bracket (or set of 4 brackets)
- 2 x Foot Base
- 16 x Miniature Ball Radial Bearing
- 17 x Metal Servo Horn Flange
- Lots of screws and nuts
- 9 x 120 oz inch MG995 or MG996 Servos
- 8 x 220 oz inch MG958 servos for the legs
- 2 x Robot hands
- Arduino Uno and Servo Shield

The next picture is the front view of the assembled 17 DOF humanoid robot configuration.

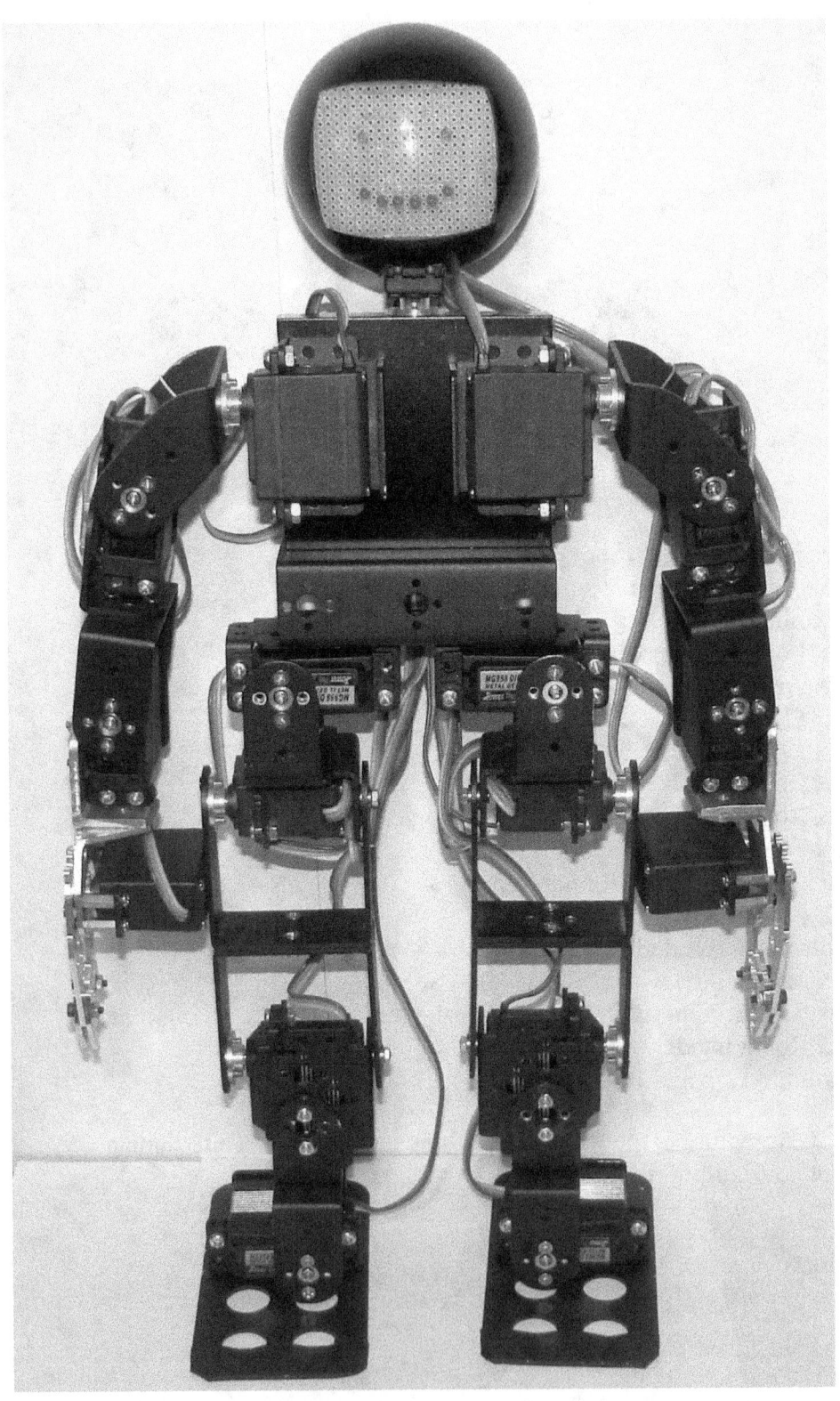

This next picture is of the back view of the assembled robot. The Arduino Uno has been mounted in place for this picture. I used three short plastic spacers between the metal and the Arduino. Only one spacer lined up with an existing screw. The other two spacers are just glued in place. Two of the servos needed four inch extensions to reach the servo shield.

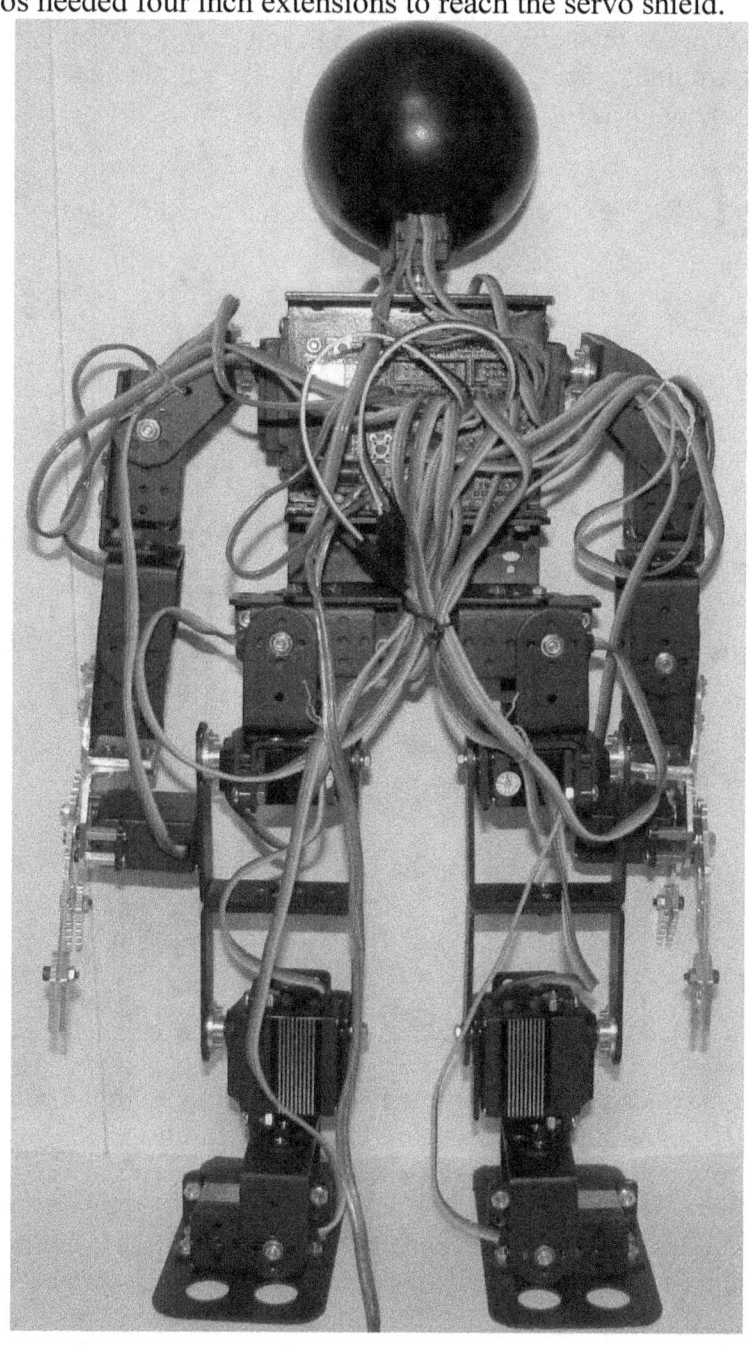

Here is a picture of the back of an earlier version with added text showing the numbers of the servos. This is so that you know how to connect the servo motors. Servo one goes to D2 and that is the third servo connector from the bottom on the servo shield. It is best to not use D0 and D1 as they are used to talk to the Arduino to update software. If you connect a servo to D0 or D1 the servo will go through some strange contortions when the Arduino is being updated. Servo 13-16 and the head (SERVO 17) go to analog 0 (D14) through analog 5 (D18).

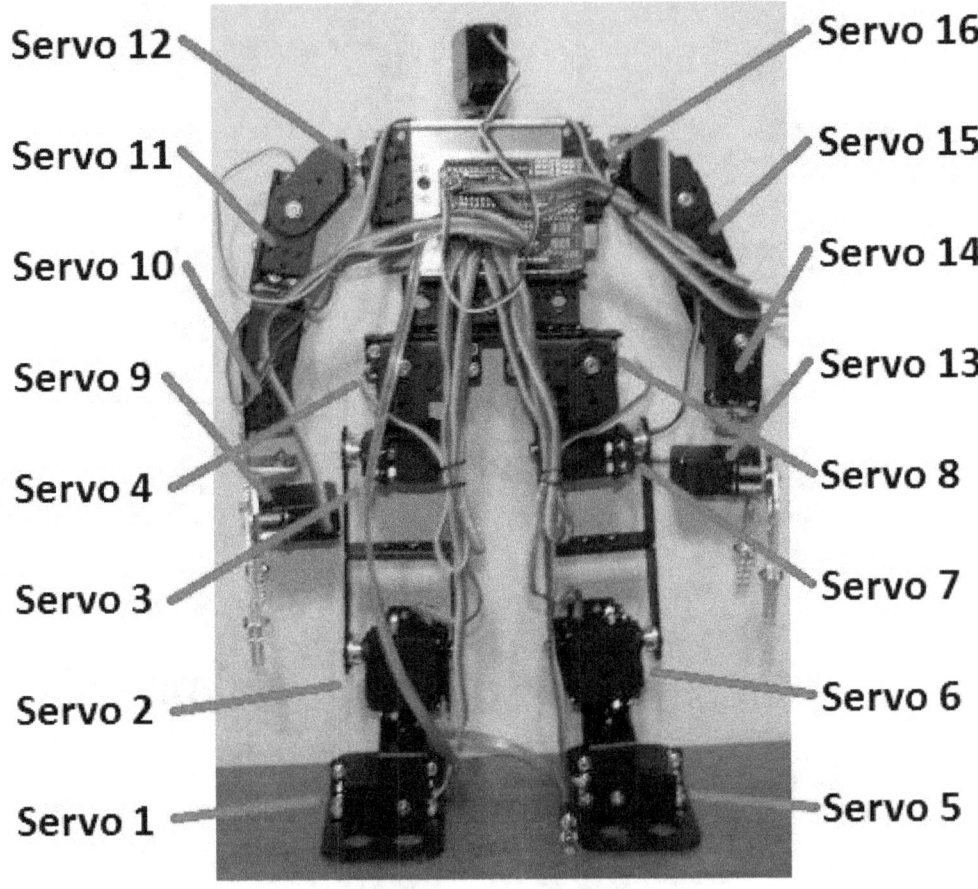

The robot may need to be supported by a strap going up to the ceiling to keep him from falling over when you turn him off. I have become used to grabbing him whenever I do a software update.

Once you update the arms, legs and chest, the next step is to calibrate the servos and then to put him back together again. With this project we are now way past the 12 servo limit. I decided to use the servo software option as the modification to servo.h is not an "authorized" change. However servo software changed the 90 degree position of all servos by about 2-3 degrees. Also you cannot use any delays or the servos will shut down. So I went back to using servo, but with the modification to servo.h that is covered in a previous chapter that allows for more than 12 servos per timer.

```
// Humanoid 17 DOF Serial Version 2
// Jan 2016 by Bob Davis
// Mapped as D2 is servo1
// Mapped as D3 is servo2
// Etc.
// f=forward, b=back
// r=right l=left
// w=wave s=sit
// c=clap p=pick up

#include <Servo.h>
// Define our Servo's
Servo servo1; // Left Ankle
Servo servo2; // Left Knee
Servo servo3; // Left Hip F-B
Servo servo4; // Left Hip R-L
Servo servo5; // Right Ankle
Servo servo6; // Right Knee
Servo servo7; // Right Hip F-B
Servo servo8; // Right Hip R-L
Servo servo9; // Left Hand
Servo servo10; // Left Wrist
Servo servo11; // Left Elbow
Servo servo12; // Left Shoulder
Servo servo13; // Right Hand
Servo servo14; // Right Wrist
Servo servo15; // Right Elbow
Servo servo16; // Right Shoulder
Servo servo17; // Head
int twait=100; // Sets time delay between steps

int walkf[8][10] = {
// 0   1    2    3    4    5   6   7   8   9
{ 95, 105, 105, 105, 105,  95, 90, 90, 85, 85}, // Left Ankle
```

```
  { 90, 90, 90, 85, 80, 75, 70, 70, 75, 80}, // Knee
  { 90, 90, 90, 85, 80, 75, 70, 70, 75, 80}, // Hip
  { 95, 100, 100, 100, 100, 95, 90, 90, 85, 90}, // Hip
  { 95, 105, 105, 105, 105, 95, 90, 90, 85, 90}, // Right Ankle
  { 90, 90, 90, 85, 80, 75, 70, 70, 75, 80}, // Knee
  { 90, 90, 90, 85, 80, 75, 70, 70, 75, 80}, // Hip
  { 95, 100, 100, 100, 100, 95, 90, 90, 85, 90}, // Hip
};
int walkb[8][10] = {
//  0   1   2   3   4   5   6   7   8   9
  { 95, 105, 105, 105, 105, 95, 90, 90, 85, 85}, // Left Ankle
  { 90, 90, 90, 95, 100, 105, 110, 110, 105, 100}, // Knee
  { 90, 90, 90, 95, 100, 105, 110, 110, 105, 100}, // Hip
  { 95, 100, 100, 100, 100, 95, 90, 90, 85, 90}, // Hip
  { 95, 105, 105, 105, 105, 95, 90, 90, 85, 90}, // Right Ankle
  { 90, 90, 90, 95, 100, 105, 110, 110, 105, 100}, // Knee
  { 90, 90, 90, 95, 100, 105, 110, 110, 105, 100}, // Hip
  { 95, 100, 100, 100, 100, 95, 90, 90, 85, 90}, // Hip
};
// Turning is a bit tricky because he has to drag his feet
int turnr[8][10] = {
//  0   1   2   3   4   5   6   7   8   9
  { 95, 100, 100, 100, 100, 95, 90, 90, 90, 90}, // Left Ankle
  { 90, 90, 90, 85, 80, 75, 70, 70, 75, 80}, // Knee
  { 90, 90, 90, 85, 80, 75, 70, 70, 75, 80}, // Hip
  { 95, 100, 100, 100, 100, 95, 90, 90, 90, 90}, // Hip
  { 95, 100, 100, 100, 100, 95, 90, 90, 90, 90}, // Right Ankle
  { 90, 90, 90, 85, 80, 75, 70, 75, 80, 85}, // Knee
  { 90, 90, 90, 85, 80, 75, 70, 75, 80, 85}, // Hip
  { 95, 100, 100, 100, 100, 95, 90, 90, 90, 90}, // Hip
};
int turnl[8][10] = {
//  0   1   2   3   4   5   6   7   8   9
  { 85, 80, 80, 80, 80, 85, 90, 90, 90, 90}, // Left Ankle
  { 90, 90, 90, 95, 100, 105, 110, 110, 105, 100}, // Knee
  { 90, 90, 90, 95, 100, 105, 110, 110, 105, 100}, // Hip
  { 85, 80, 80, 80, 80, 85, 90, 90, 90, 90}, // Hip
  { 85, 80, 80, 80, 80, 85, 90, 90, 90, 90}, // Right Ankle
  { 90, 90, 90, 95, 100, 105, 110, 105, 100, 95}, // Knee
  { 90, 90, 90, 95, 100, 105, 110, 105, 100, 95}, // Hip
  { 85, 80, 80, 80, 80, 85, 90, 90, 90, 90}, // Hip
```

```
};

void setup(){
 servo1.attach(2); // servo on digital pin 2
 servo2.attach(3); // servo on digital pin 3
 servo3.attach(4); // servo on digital pin 4
 servo4.attach(5); // servo on digital pin 5
 servo5.attach(6); // servo on digital pin 6
 servo6.attach(7); // servo on digital pin 7
 servo7.attach(8); // servo on digital pin 8
 servo8.attach(9); // servo on digital pin 9
 servo9.attach(10); // servo on digital pin 10
 servo10.attach(11); // servo on digital pin 11
 servo11.attach(12); // servo on digital pin 12
 servo12.attach(13); // servo on digital pin 13
 servo13.attach(14); // servo on digital pin 14
 servo14.attach(15); // servo on digital pin 15
 servo15.attach(16); // servo on digital pin 16
 servo16.attach(17); // servo on digital pin 17
 servo17.attach(18); // servo on digital pin 18
 Serial.begin(9600);
 Serial.print("Ready");
}
void loop(){
 servo1.write(90); // Return to zero position
 servo2.write(90);
 servo3.write(90);
 servo4.write(90);
 servo5.write(90);
 servo6.write(90);
 servo7.write(90);
 servo8.write(90);
 servo9.write(90);
 servo10.write(90);
 servo11.write(90);
 servo12.write(160);// 160=Left Shoulder down
 servo13.write(90);
 servo14.write(90);
 servo15.write(90);
 servo16.write(20); // 20=Right Shoulder down
 servo17.write(90); // Head
```

```
if ( Serial.available()) {
  char sinch = Serial.read();
  // Test out the servos 2 at a time
  switch(sinch) {
    case '1':
      servo1.write(60); //1=Feet
      servo5.write(60);
      delay(twait*2);
      servo1.write(120);
      servo5.write(120);
      delay(twait*2);
      break;
    case '2':
      servo2.write(60); //2=Knees
      servo6.write(60);
      delay(twait*2);
      servo2.write(120);
      servo6.write(120);
      delay(twait*2);
      break;
    case '3':
      servo3.write(60); //3=Hips
      servo7.write(60);
      delay(twait*2);
      servo3.write(120);
      servo7.write(120);
      delay(twait*2);
      break;
    case '4':
      servo4.write(60); //4=Hips
      servo8.write(60);
      delay(twait*2);
      servo4.write(120);
      servo8.write(120);
      delay(twait*2);
      break;
    case '5':
      servo9.write(60); //5=hand
      servo13.write(60);
      delay(twait*2);
```

```
      servo9.write(120);
      servo13.write(120);
      delay(twait*2);
      break;
    case '6':
      servo10.write(60); //6=wrist
      servo14.write(60);
      delay(twait*2);
      servo10.write(120);
      servo14.write(120);
      delay(twait*2);
      break;
    case '7':
      servo11.write(60); //7=elbow
      servo15.write(60);
      delay(twait*2);
      servo11.write(120);
      servo15.write(120);
      delay(twait*2);
      break;
    case '8':
      servo12.write(90); //8=Shoulders
      servo16.write(90);
      delay(twait*2);
      break;
    case '9':
      servo17.write(60); //9=Head
      delay(twait*2);
      servo17.write(120);
      delay(twait*2);
      break;
    case 's': // sit down
      servo2.write(35);
      servo3.write(160);
      servo6.write(145);
      servo7.write(20);
      delay(twait*10);
      servo2.write(30); // lean forward to stand back up
      servo3.write(180);
      servo6.write(150);
      servo7.write(00);
```

```
        delay(twait*10);
        break;
    case 'w': // wave hand
        servo12.write(30);
        servo10.write(100);
        delay(twait*5);
        servo10.write(160);
        delay(twait*5);
        servo10.write(100);
        delay(twait*5);
        servo10.write(160);
        delay(twait*5);
        servo10.write(100);
        delay(twait*5);
        break;
    case 'p': // pick up
        servo10.write(110);  // hands down
        servo14.write(70);
        servo12.write(140); // arms forward
        servo16.write(40);
        servo2.write(70); // bend down
        servo3.write(145);
        servo6.write(110);
        servo7.write(35);
        delay(twait*5);
        servo9.write(60); // close hands
        servo13.write(60);
        delay(twait*5);
        servo2.write(70); // rise up slowly
        servo3.write(120);
        servo6.write(110);
        servo7.write(60);
        delay(twait*2);
        servo2.write(85); // rise up
        servo3.write(90);
        servo6.write(95);
        servo7.write(90);
        delay(twait*5);
        servo12.write(90); // raise arms
        servo16.write(90);
        delay(twait*10);
```

```cpp
    break;
  case 'c': // Clap hands
    servo2.write(85); // Lean back to counterbalance arms
    servo6.write(95);
    servo9.write(90); // move arms forward
    servo10.write(120);
    servo11.write(120);
    servo12.write(90);
    servo13.write(90);
    servo14.write(60);
    servo15.write(60);
    servo16.write(90);
    servo17.write(40);
    delay(twait*5);
    servo9.write(60); //pinch hands
    servo13.write(60);
    servo17.write(140);
    delay(twait*5);
    servo9.write(90); //open hands
    servo13.write(90);
    servo17.write(40);
    delay(twait*5);
    break;
  // Sequential operations
  case 'r': // right key;
    for (int i=0; i<10; i++){
      servo1.write(turnr[0][i]);
      servo2.write(turnr[1][i]);
      servo3.write(turnr[2][i]);
      servo4.write(turnr[3][i]);
      servo5.write(turnr[4][i]);
      servo6.write(turnr[5][i]);
      servo7.write(turnr[6][i]);
      servo8.write(turnr[7][i]);
      delay(twait);
    }
    break;
  case 'b': // back key
    for (int i=0; i<10; i++){
      servo1.write(walkb[0][i]);
      servo2.write(walkb[1][i]);
```

```cpp
        servo3.write(walkb[2][i]);
        servo4.write(walkb[3][i]);
        servo5.write(walkb[4][i]);
        servo6.write(walkb[5][i]);
        servo7.write(walkb[6][i]);
        servo8.write(walkb[7][i]);
        delay(twait);
      }
      break;
    case 'l': // left key;
      for (int i=0; i<10; i++){
        servo1.write(turnl[0][i]);
        servo2.write(turnl[1][i]);
        servo3.write(turnl[2][i]);
        servo4.write(turnl[3][i]);
        servo5.write(turnl[4][i]);
        servo6.write(turnl[5][i]);
        servo7.write(turnl[6][i]);
        servo8.write(turnl[7][i]);
        delay(twait);
      }
      break;
    case 'f': // forward key
      for (int i=0; i<10; i++){
        servo1.write(walkf[0][i]);
        servo2.write(walkf[1][i]);
        servo3.write(walkf[2][i]);
        servo4.write(walkf[3][i]);
        servo5.write(walkf[4][i]);
        servo6.write(walkf[5][i]);
        servo7.write(walkf[6][i]);
        servo8.write(walkf[7][i]);
        delay(twait);
      }
      break;
  }
 }
} // End of program
```

Chapter 8

Walking Dog Robot

14 DOF

This next project is to make a four legged walking robot or a "Dog". You could also call it a "horse". Getting it to walk is not as easy as it seems. Sometimes one of the legs will not even touch the ground. That has an effect on the amount of movement that leg can create.

The dogs body is 7.5 inches long. It had to be long so its legs do not hit each other. It can be made with four waist brackets or two waist brackets and the home made chest. I used four of the straight brackets to space the back two waist brackets further apart.

The neck is made out of two of the angled U shaped brackets. That moves the head both out and up at the same time. The head is a hand turned sideways. A "L" shaped bracket would have been ideal but I had used all four of mine. I used a hardware store "L" bracket instead.

The legs are made about the same way as for the 13 DOF human but the ankles are turned the other way. This change gives three joints all moving in the same direction. This then makes it easier to lift a foot straight up. I had to remove his "feet" as there was not enough room for them as they actually hit each other front to back. He also had no need for those large feet as he is four legged. Here is a close up picture of his back legs showing how they go together.

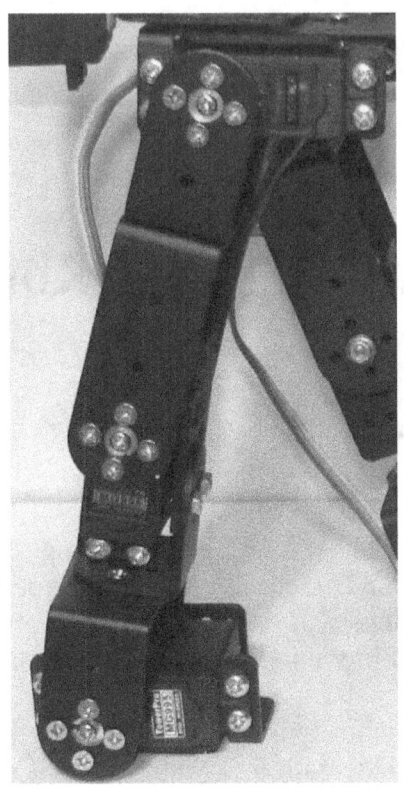

This is a picture of the completed dog robot.

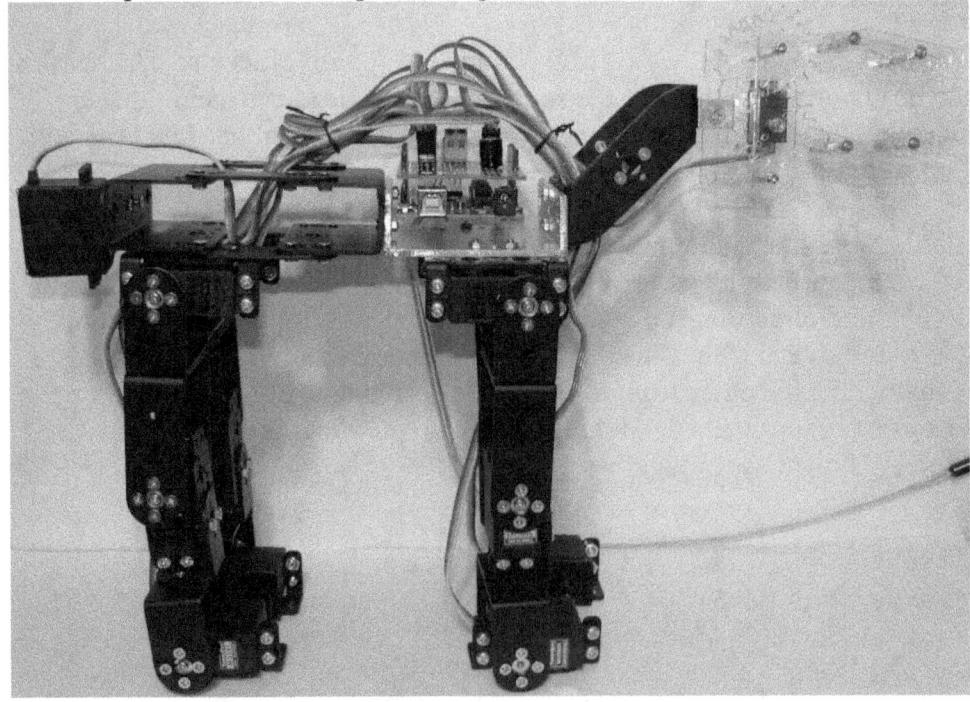

This picture of the dog robot has the servo numbers labeled.

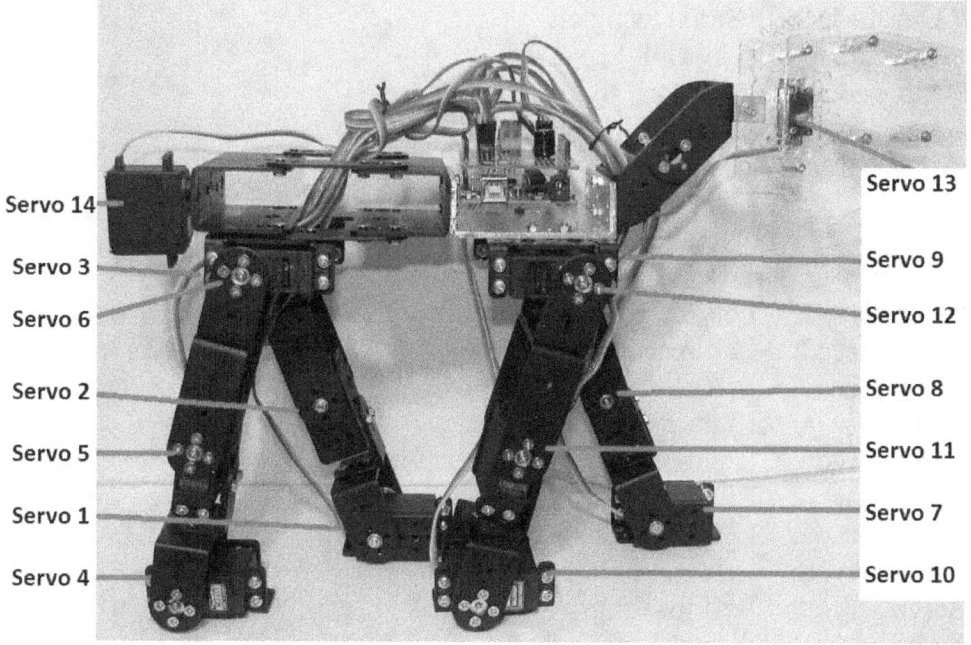

This is the 14 DOF Dog robot parts list:

- 12 x Multi-functional servo mounting bracket
- 8 x Short U-type servo bracket
- 4 x Long U-type servo bracket
- 2 x Angled U-type servo bracket
- 4 x Flat servo bracket
- 5 x L-type servo bracket
- 2 x Robot waist bracket
- 1 x Robot chest bracket (or set of 2 waist brackets)
- 0 x Foot Base (or you can make smaller ones)
- 12 x Miniature Ball Radial Bearing
- 13 x Metal Servo Horn Flange
- Lots of screws and nuts
- 13 x MG995 or MG996 Servos
- 1 x Robot hand (As the head)
- Arduino Uno and Servo Shield

Coming up next is the code to make the dog work. He has "f" for forward, "b" for back "r" for right "l" for left and "s" for sit as his commands. We are now using 12x12 arrays to run 12 servos through 12 steps. The array for sitting has two more servos added for the head and tail.

```
// Dog 14 DOF + Serial using arrays
// November 2015 by Bob Davis
// Mapped as D2 is servo1
// Mapped as D3 is servo2
// Etc.

#include <Servo.h>
// Define our Servo's
Servo servo1; // Left Rear Ankle - needs extension
Servo servo2; // Left rear Knee
Servo servo3; // Left rear hip
Servo servo4; // Right rear Ankle
Servo servo5; // Right rear knee
Servo servo6; // Right rear hip
Servo servo7; // Left Front ankle
Servo servo8; // Left front knee
Servo servo9; // Left front hip
Servo servo10; // Right Front ankle
Servo servo11; // Right front knee
Servo servo12; // Right front hip
Servo servo13; // Head
Servo servo14; // Tail
int twait=100; // Sets time delay between steps

int walkf[12][12] = {
// 0   1    2    3    4    5    6    7    8    9   10   11
{ 90,  90,  90,  90, 100, 110, 100,  90,  90,  90,  90,  90}, // L Rear Ankle
{ 90,  90,  90,  80,  70,  60,  70,  80,  90,  90,  90,  90}, // Knee
{ 90, 100, 100,  80,  70,  60,  70,  80,  80,  90,  90,  90}, // Hip
{ 90,  90,  90,  90,  90,  90,  90,  90,  90,  80,  70,  80}, // R Rear ankle
{ 90,  90,  90,  90,  90,  90,  90,  90,  90, 100, 110,  90}, // Knee
{ 90,  80,  80,  80,  80,  80,  80,  80,  80, 100, 110,  90}, // hip
{ 90,  90,  90,  90,  90,  90,  90, 100, 110, 100,  90,  90}, // L Front ankle
{ 90,  90,  90,  90,  90,  90,  80,  70,  60,  70,  80,  90}, // Knee
{ 90, 100, 100, 100, 100, 100,  80,  70,  60,  70,  80,  90}, // hip
{ 90,  80,  70,  80,  90,  90,  90,  90,  90,  90,  90,  90}, // R Front ankle
{100, 110, 120, 110, 100,  90,  90,  90,  90,  90,  90,  90}, // Knee
{100, 110, 120, 110, 100,  90,  90,  90,  90,  90,  90,  90}, // Hip
};
```

```c
int walkb[12][12] = {
//  0   1    2    3    4    5    6    7    8    9   10  11
{ 90,  90,  90,  90,  90,  90,  90,  90,  90,  80,  70, 80}, // L Rear ankle
{ 90,  90,  90,  90,  90,  90,  90,  90,  90, 100, 110, 90}, // Knee
{ 90,  80,  80,  80,  80,  80,  80,  80,  80, 100, 110, 90}, // hip
{ 90,  90,  90, 100, 110, 120, 110, 100,  90,  90,  90, 90}, // R Rear Ankle
{ 90,  90,  90,  80,  70,  60,  70,  80,  90,  90,  90, 90}, // Knee
{ 90, 100, 100,  80,  70,  60,  70,  80,  80,  90,  90, 90}, // Hip
{ 80,  70,  60,  70,  80,  90,  90,  90,  90,  90,  90, 90}, // L Front ankle
{100, 110, 120, 110, 100,  90,  90,  90,  90,  90,  90, 90}, // Knee
{100, 110, 120, 110, 100,  90,  90,  90,  90,  90,  90, 90}, // Hip
{ 90,  90,  90,  90,  90,  90, 100, 110, 120, 110, 100, 90}, // R Front ankle
{ 90,  90,  90,  90,  90,  90,  80,  70,  60,  70,  90, 90}, // Knee
{ 90, 100, 100, 100, 100, 100,  90,  80,  70,  80,  90, 90}, // hip
};

// Turning is a bit tricky he digs in his feet
int turnr[12][12] = {
//  0   1    2    3    4    5    6    7    8    9
{ 90,  90,  90,  90,  90,  90,  90,  90,  90,  90,  90, 90}, // Left Rear Ankle
{ 90,  90,  90,  90,  90,  90,  90,  90,  90,  90,  90, 90}, // Knee
{ 90,  90,  90,  90,  90,  90,  90,  90,  90,  90,  90, 90}, // Hip
{ 90,  90,  90,  90,  90,  90, 110, 120, 120, 120, 110, 100}, // R Rear ankle
{ 90,  90,  90,  90,  90,  90, 100, 110, 120, 110, 100,  90}, // Knee
{ 90,  90,  90,  90,  90,  90, 100, 110, 120, 110, 100,  90}, // hip
{ 90,  90,  90,  90,  90,  90,  90,  90,  90,  90,  90, 90}, // Left Front ankle
{ 90,  90,  90,  90,  90,  90,  90,  90,  90,  90,  90, 90}, // Knee
{ 90,  90,  90,  90,  90,  90,  90,  90,  90,  90,  90, 90}, // hip
{110, 120, 120, 120, 110, 100,  90,  90,  90,  90,  90, 90}, // R Front ankle
{100, 110, 120, 110, 100,  90,  90,  90,  90,  90,  90, 90}, // Knee
{100, 110, 120, 110, 100,  90,  90,  90,  90,  90,  90, 90}, // Hip
};

int turnl[12][12] = {
//  0   1    2    3    4    5    6    7    8    9
{ 90,  90,  90,  80,  70,  60,  70,  80,  90,  90,  90, 90}, // Left Rear Ankle
{ 90,  90,  90,  80,  70,  60,  70,  80,  90,  90,  90, 90}, // Knee
{ 90, 100, 100,  80,  70,  60,  70,  80,  80,  90,  90, 90}, // Hip
{ 90,  90,  90,  90,  90,  90,  90,  90,  90,  90,  90, 90}, // Right Rear ankle
{ 90,  90,  90,  90,  90,  90,  90,  90,  90,  90,  90, 90}, // Knee
{ 90,  90,  90,  90,  90,  90,  90,  90,  90,  90,  90, 90}, // hip
```

```cpp
{ 90,  90,  90,  90,  90,  90,  80,  70,  60,  70,  90,  90}, // Left Front ankle
{ 90,  90,  90,  90,  90,  90,  80,  70,  60,  70,  90,  90}, // Knee
{ 90, 100, 100, 100, 100, 100,  90,  80,  70,  80,  90,  90}, // hip
{ 90,  90,  90,  90,  90,  90,  90,  90,  90,  90,  90,  90}, // Right Front ankle
{ 90,  90,  90,  90,  90,  90,  90,  90,  90,  90,  90,  90}, // Knee
{ 90,  90,  90,  90,  90,  90,  90,  90,  90,  90,  90,  90}, // Hip
};

int clap[14][12] = {  //now sit
// 0   1   2   3   4   5   6   7   8   9
{ 95, 100, 105, 110, 115, 120, 125, 120, 115, 110, 105,100},//L Rear ankle
{ 80,  70,  60,  50,  40,  30,  20,  30,  40,  50,  60,  70}, // Knee
{ 85,  80,  75,  70,  65,  60,  55,  60,  65,  70,  75,  80}, // Hip
{ 85,  80,  75,  70,  65,  60,  55,  60,  65,  70,  75,  80}, // Right Rear Ankle
{100, 110, 120, 130, 140, 150, 160, 150, 140, 130, 120, 110}, // Knee
{ 95, 100, 105, 110, 115, 120, 125, 120, 115, 110, 105, 100}, // hip
{ 90,  80,  70,  60,  50,  30,  10,  30,  50,  60,  70,  80}, // Left Front ankle
{ 90,  90,  90,  90,  90,  90,  90,  90,  90,  90,  90,  90}, // Knee
{ 90,  90,  90,  90,  90,  90,  90,  90,  90,  90,  90,  90}, // hip
{ 90, 100, 110, 120, 130, 150, 170, 150, 130,120,110,100},//R Front ankle
{ 90,  90,  90,  90,  90,  90,  90,  90,  90,  90,  90,  90}, // Knee
{ 90,  90,  90,  90,  90,  90,  90,  90,  90,  90,  90,  90}, // Hip
{ 90,  90,  90,  90,  90,  60,  60,  60,  90,  90,  90,  90}, // head
{ 90,  90,  90, 130,  90,  50,  90, 130,  90,  50,  90,  90}, // Tail
};

void setup(){
  servo1.attach(2); // servo on digital pin 2
  servo2.attach(3); // servo on digital pin 3
  servo3.attach(4); // servo on digital pin 4
  servo4.attach(5); // servo on digital pin 5
  servo5.attach(6); // servo on digital pin 6
  servo6.attach(7); // servo on digital pin 7
  servo7.attach(8); // servo on digital pin 8
  servo8.attach(9); // servo on digital pin 9
  servo9.attach(10); // servo on digital pin 10
  servo10.attach(11); // servo on digital pin 11
  servo11.attach(12); // servo on digital pin 12
  servo12.attach(13); // servo on digital pin 13
  servo13.attach(14); // servo on digital pin 10
  servo14.attach(15); // servo on digital pin 11
```

```
  Serial.begin(9600);
  Serial.print("Ready");
}

void loop(){
  servo1.write(90); // Return to zero position
  servo2.write(90);
  servo3.write(90);
  servo4.write(90);
  servo5.write(90);
  servo6.write(90);
  servo7.write(90);
  servo8.write(90);
  servo9.write(90);
  servo10.write(90);
  servo11.write(90);
  servo12.write(90);
  servo13.write(90);
  servo14.write(90);
  if ( Serial.available()) {
    char sinch = Serial.read();
    // Test out the servos 4 at a time
    switch(sinch) {
      case '1': //1 Key - Ankles
        servo1.write(50);
        servo4.write(50);
        servo7.write(50);
        servo10.write(50);
        delay(twait*2);
        servo1.write(130);
        servo4.write(130);
        servo7.write(130);
        servo10.write(130);
        delay(twait*2);
        break;
      case '2': //2 Key - Knees
        servo2.write(50);
        servo5.write(50);
        servo8.write(50);
        servo11.write(50);
        delay(twait*2);
```

```cpp
      servo2.write(130);
      servo5.write(130);
      servo8.write(130);
      servo11.write(130);
      delay(twait*2);
      break;
    case '3': //3 Key - Hips
      servo3.write(50);
      servo6.write(50);
      servo9.write(50);
      servo12.write(50);
      delay(twait*2);
      servo3.write(130);
      servo6.write(130);
      servo9.write(130);
      servo12.write(130);
      delay(twait*2);
      break;
    case '4': //4 Key - Head
      servo13.write(50);
      delay(twait*2);
      servo13.write(130);
      delay(twait*2);
      break;
    case '5': //5 Key - Tail
      servo14.write(50);
      delay(twait*2);
      servo14.write(130);
      delay(twait*2);
      break;
    // Sequential operations
    case 's': // Sit - Sit and shake head and tail
      for (int i=0; i<12; i++){
        servo1.write(clap[0][i]);
        servo2.write(clap[1][i]);
        servo3.write(clap[2][i]);
        servo4.write(clap[3][i]);
        servo5.write(clap[4][i]);
        servo6.write(clap[5][i]);
        servo7.write(clap[6][i]);
        servo8.write(clap[7][i]);
```

```
          servo9.write(clap[8][i]);
          servo10.write(clap[9][i]);
          servo11.write(clap[10][i]);
          servo12.write(clap[11][i]);
          servo13.write(clap[12][i]);
          servo14.write(clap[13][i]);
          delay(twait*2);
        }
        break;
      case 'r': // right key;
        for (int i=0; i<12; i++){
          servo1.write(turnr[0][i]);
          servo2.write(turnr[1][i]);
          servo3.write(turnr[2][i]);
          servo4.write(turnr[3][i]);
          servo5.write(turnr[4][i]);
          servo6.write(turnr[5][i]);
          servo7.write(turnr[5][i]);
          servo8.write(turnr[6][i]);
          servo9.write(turnr[8][i]);
          servo10.write(turnr[9][i]);
          servo11.write(turnr[10][i]);
          servo12.write(turnr[11][i]);
          delay(twait);
        }
        break;
      case 'b':  // back key
        for (int i=0; i<12; i++){
          servo1.write(walkb[0][i]);
          servo2.write(walkb[1][i]);
          servo3.write(walkb[2][i]);
          servo4.write(walkb[3][i]);
          servo5.write(walkb[4][i]);
          servo6.write(walkb[5][i]);
          servo7.write(walkb[6][i]);
          servo8.write(walkb[7][i]);
          servo9.write(walkb[8][i]);
          servo10.write(walkb[9][i]);
          servo11.write(walkb[10][i]);
          servo12.write(walkb[11][i]);
          delay(twait);
```

```
        }
      break;
    case 'l': // left key;
      for (int i=0; i<12; i++){
        servo1.write(turnl[0][i]);
        servo2.write(turnl[1][i]);
        servo3.write(turnl[2][i]);
        servo4.write(turnl[3][i]);
        servo5.write(turnl[4][i]);
        servo6.write(turnl[5][i]);
        servo7.write(turnl[6][i]);
        servo8.write(turnl[7][i]);
        servo9.write(turnl[8][i]);
        servo10.write(turnl[9][i]);
        servo11.write(turnl[10][i]);
        servo12.write(turnl[11][i]);
        delay(twait);
      }
      break;
    case 'f': // forward key
      for (int i=0; i<12; i++){
        servo1.write(walkf[0][i]);
        servo2.write(walkf[1][i]);
        servo3.write(walkf[2][i]);
        servo4.write(walkf[3][i]);
        servo5.write(walkf[4][i]);
        servo6.write(walkf[5][i]);
        servo7.write(walkf[6][i]);
        servo8.write(walkf[7][i]);
        servo9.write(walkf[8][i]);
        servo10.write(walkf[9][i]);
        servo11.write(walkf[10][i]);
        servo12.write(walkf[11][i]);
        delay(twait);
      }
      break;
    }
  }
} // End of program
```

Chapter 9

Walking Dinosaur Robot

17 DOF

The dinosaur robot is the dog robot with a long tail attached. Just like in the humanoid robots this project builds on the previous project. The tail adds 3 more servos to the project bringing the total to 17 servos. I did not have any more straight brackets so I used some metal hardware like what you would find at the hardware store. Because of a servo shortage there is at least one Hitec HS-422 servo used in the tail. These servos come with a plastic horn and do not fit the metal horns. The plastic horns have smaller holes in them so you will have to use smaller screws to fasten the U shaped brackets to the horns.

The long tail will also require some servo cable extensions. I made three of the extension cables but one was only about three inches long. The other two cables were about 10 inches long. Two of the extensions were made out of servo cables taken from some burned up servos.

Here is a close up picture of the long dinosaur tail.

I also sanded and painted the moving parts of his "mouth" black. The clear plastic mouth was not scary enough. The next picture shows the complete dinosaur robot. Adding the new tail will more than double his length. From head to tail he is now about 30 inches long.

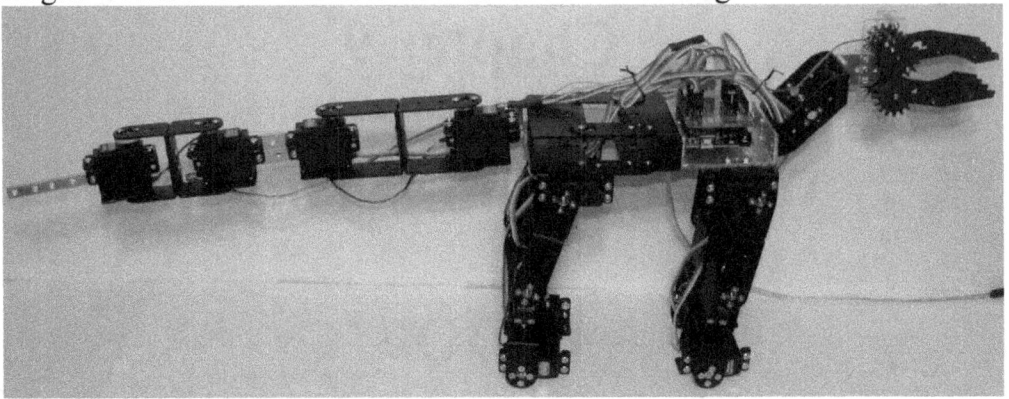

Here is a front view showing his optional LED eyes and a longer neck.

This is the updated 17 DOF Dinosaur robot parts list:

- 16 x Multi-functional servo mounting bracket
- 10 x Short U-type servo bracket
- 6 x Long U-type servo bracket
- 4 x Angled U-type servo bracket (4 for a longer neck)
- 6 x Flat servo bracket
- 6 x L-type servo bracket
- 2 x Robot waist bracket
- 1 x Robot chest bracket (or set of 2 waist brackets)
- 0 x Foot Base (or you can make smaller ones)
- 16 x Miniature Ball Radial Bearing
- 16 x Metal Servo Horn Flange
- Lots of screws and nuts
- 16 x MG995 or MG996 Servos
- 1 x Robot hand (As the head, includes a micro servo)
- Arduino Uno and Servo Shield

For the dinosaur software all we need to do is to add the "t" for tail function to the dog's software, right? Wrong! The weight of the tail messed everything up. He can no longer lift his rear legs because of the added weight. So to walk and turn he had to really dig in his heels and move his tail to lift the rear legs. Also if he wags his tail while sitting he falls over. If he leaves his tail straight while sitting the tail touches the ground and keeps him from falling over. He now has a "t" for tail but stand clear! His tail really flies, and that is at only 30 degrees per servo. It will send anything that gets in its way flying as well!

```
// Dinosaur 17 DOF Ver 2 with Serial using arrays
// November 2015 by Bob Davis
// Mapped as D2 is servo1
// Mapped as D3 is servo2
// Etc.
// Available Commands:
// 0=zero servos, 1=ankles, 2=knees, 3=hips
// f=forward, b=back, r=right, l=left
// s=sit, d=down, u=up
// if no command wiggles tail and mouth

#include <Servo.h>
```

```
// Define our Servo's
Servo servo1; // Left Rear Ankle - needs extension
Servo servo2; // Left rear Knee
Servo servo3; // Left rear hip
Servo servo4; // Right rear Ankle
Servo servo5; // Right rear knee
Servo servo6; // Right rear hip
Servo servo7; // Left Front ankle
Servo servo8; // Left front knee
Servo servo9; // Left front hip
Servo servo10; // Right Front ankle
Servo servo11; // Right front knee
Servo servo12; // Right front hip
Servo servo13; // Head
Servo servo14; // Tail 1
Servo servo15; // Tail 2
Servo servo16; // Tail 3
Servo servo17; // Tail 4
int twait=100; // Sets time delay between steps

int walkf[16][12] = {
// 0  1   2   3   4   5   6   7   8   9  10  11
{ 90, 90,100,100,110,100,100, 90, 90, 90, 90, 90}, // L R Ankle
{ 90, 90, 80, 70, 60, 70, 80, 90, 90, 90, 90, 90}, // Knee
{ 90, 90, 80, 70, 60, 70, 80, 80, 80, 80, 90, 90}, // Hip
{ 90, 90, 90, 90, 90, 90, 90, 90, 90, 80, 70, 90}, // R R ankle
{ 90, 90, 90, 90, 90, 90, 90, 90, 90,100,110, 90}, // Knee
{ 90, 90, 80, 80, 80, 80, 80, 80, 80,100,110, 90}, // hip
{ 90, 90, 90, 90, 90, 90,100,110,120,120,110, 90}, // L F ankle
{ 90, 90, 90, 90, 90, 90, 80, 70, 60, 70, 80, 90}, // Knee
{ 90, 90,100,100,100,100, 80, 70, 60, 60, 70, 90}, // hip
{ 90, 80, 70, 60, 70, 80, 90, 90, 90, 90, 90, 90}, // R F ankle
{100,110,120,110,100, 90, 90, 90, 90, 90, 90, 90}, // Knee
{100,110,120,110,100,100,100,100,100,100, 90, 90}, // Hip
{100,100,100,100, 90, 90, 80, 80, 80, 80, 90, 90}, // Tail1
{ 80, 80, 80, 80, 90, 90,100,100,100,100, 90, 90}, // Tail2
{100,100,100,100, 90, 90, 80, 80, 80, 80, 90, 90}, // Tail3
{ 90, 90, 90, 90, 90, 90, 90, 90, 90, 90, 90, 90}, // Tail4
};

int walkb[16][12] = {
```

```
//0   1   2   3   4   5   6   7   8   9  10  11
{ 90, 90, 90, 90, 90, 90, 90, 90, 90, 80, 70, 90}, // L R ankle
{ 90, 90, 90, 90, 90, 90, 90, 90, 90,100,100, 90}, // Knee
{ 90, 80, 80, 80, 80, 80, 80, 80, 80,100,100, 90}, // hip
{ 90, 90,100,110,120,110,100, 90, 90, 90, 90, 90}, // R R Ankle
{ 90, 90, 80, 70, 60, 70, 80, 90, 90, 90, 90, 90}, // Knee
{ 90, 90, 80, 70, 60, 70, 80, 80, 90, 90, 90, 90}, // Hip
{ 80, 70, 60, 60, 70, 80, 90, 90, 90, 90, 90, 90}, // L F ankle
{100,110,120,110,100, 90, 90, 90, 90, 90, 90, 90}, // Knee
{100,110,120,110,100, 90, 90, 90, 90, 90, 90, 90}, // Hip
{ 90, 90, 90, 90, 90, 90,100,110,120,120,110, 90}, // R F ankle
{ 90, 90, 90, 90, 90, 90, 80, 70, 60, 70, 90, 90}, // Knee
{ 90,100,100,100,100,100, 90, 80, 70, 80, 90, 90}, // hip
{ 80, 80, 80, 80, 90, 90,100,100,100,100, 90, 90}, // Tail1
{100,100,100,100, 90, 90, 80, 80, 80, 80, 90, 90}, // Tail2
{ 80, 80, 80, 80, 90, 90,100,100,100,100, 90, 90}, // Tail3
{ 90, 90, 90, 90, 90, 90, 90, 90, 90, 90, 90, 90}, // Tail4
};

// Turning is a bit tricky as he digs in his feet
int turnr[12][12] = {
//0   1   2   3   4   5   6   7   8   9  10  11
{ 90, 90, 90, 90, 90, 90, 90, 90, 90, 90, 90, 90}, // L R Ankle
{ 90, 90, 90, 90, 90, 90, 90, 90, 90, 90, 90, 90}, // Knee
{ 90, 90, 90, 90, 90, 90, 90, 90, 90, 90, 90, 90}, // Hip
{ 90, 90, 90, 95, 95, 95, 95, 95, 90, 90, 90, 90}, // R R ankle
{ 90, 90, 90, 95, 95, 95, 95, 95, 90, 90, 90, 90}, // Knee
{ 90, 90, 90, 95, 95, 95, 95, 95, 90, 90, 90, 90}, // hip
{ 90, 90, 90, 90, 90, 90, 90, 90, 90, 90, 90, 90}, // L F ankle
{ 90, 90, 90, 90, 90, 90, 90, 90, 90, 90, 90, 90}, // Knee
{ 90, 90, 90, 90, 90, 90, 90, 90, 90, 90, 90, 90}, // hip
{ 90, 90, 90, 90,100,110,110,125,110,100, 90, 90}, // R F ankle
{ 90, 90, 90, 90,100,110,110,110, 90, 90, 90, 90}, // Knee
{ 90, 90, 90, 90,100,100,110,100, 90, 90, 90, 90}, // Hip
};

int turnl[12][12] = {
//0   1   2   3   4   5   6   7   8   9  10  11
{ 90, 90, 90, 80, 80, 80, 80, 80, 90, 90, 90, 90}, // L R Ankle
{ 90, 90, 90, 80, 80, 80, 80, 80, 90, 90, 90, 90}, // Knee
{ 90, 90, 90, 80, 80, 80, 80, 80, 80, 90, 90, 90}, // Hip
```

```
{ 90, 90, 90, 90, 90, 90, 90, 90, 90, 90, 90, 90}, // R R ankle
{ 90, 90, 90, 90, 90, 90, 90, 90, 90, 90, 90, 90}, // Knee
{ 90, 90, 90, 90, 90, 90, 90, 90, 90, 90, 90, 90}, // hip
{ 90, 90, 90, 90, 80, 70, 70, 65, 70, 80, 90, 90}, // L F ankle
{ 90, 90, 90, 90, 80, 70, 70, 70, 90, 90, 90, 90}, // Knee
{ 90, 90, 90, 90, 90, 80, 70, 80, 90, 90, 90, 90}, // hip
{ 90, 90, 90, 90, 90, 90, 90, 90, 90, 90, 90, 90}, // R F ankle
{ 90, 90, 90, 90, 90, 90, 90, 90, 90, 90, 90, 90}, // Knee
{ 90, 90, 90, 90, 90, 90, 90, 90, 90, 90, 90, 90}, // Hip
};

int sit[14][12] = {  //now sit
// 0   1   2   3   4   5   6   7   8   9  10  11
{ 95,100,105,110,115,120,125,120,115,110,105,100}, // L R ankle
{ 80, 70, 60, 50, 40, 30, 20, 30, 40, 50, 60, 80}, // Knee
{ 85, 80, 75, 70, 65, 60, 55, 60, 65, 70, 75, 80}, // Hip
{ 85, 80, 75, 70, 65, 60, 55, 60, 65, 70, 75, 80}, // R R Ankle
{100,110,120,130,140,150,160,150,140,130,120,100}, // Knee
{ 95,100,105,110,115,120,125,120,115,110,105,100}, // hip
{ 90, 80, 70, 60, 50, 30, 10, 30, 50, 60, 70, 80}, // L F ankle
{ 90, 90, 90, 90, 90, 90, 90, 90, 90, 90, 90, 90}, // Knee
{ 90, 90, 90, 90, 90, 90, 90, 90, 90, 90, 90, 90}, // hip
{ 90,100,110,120,130,150,170,150,130,120,110,100}, // R F ankle
{ 90, 90, 90, 90, 90, 90, 90, 90, 90, 90, 90, 90}, // Knee
{ 90, 90, 90, 90, 90, 90, 90, 90, 90, 90, 90, 90}, // Hip
{ 90, 90, 90, 90, 90, 60, 60, 60, 90, 90, 90, 90}, // head
{ 90,110,130,150,170,170,170,150,120,110,100, 90}, // Tail
};

void setup(){
  servo1.attach(2); // servo on digital pin 2
  servo2.attach(3); // servo on digital pin 3
  servo3.attach(4); // servo on digital pin 4
  servo4.attach(5); // servo on digital pin 5
  servo5.attach(6); // servo on digital pin 6
  servo6.attach(7); // servo on digital pin 7
  servo7.attach(8); // servo on digital pin 8
  servo8.attach(9); // servo on digital pin 9
  servo9.attach(10); // servo on digital pin 10
  servo10.attach(11); // servo on digital pin 11
  servo11.attach(12); // servo on digital pin 12
```

```
    servo12.attach(13); // servo on digital pin 13
    servo13.attach(14); // servo on digital pin 10
    servo14.attach(15); // servo on digital pin 11
    servo15.attach(16); // servo on digital pin 11
    servo16.attach(17); // servo on digital pin 11
    servo17.attach(18); // servo on digital pin 11
    Serial.begin(9600);
    Serial.print("Ready");
}

void loop(){
  if ( Serial.available()) {
    char sinch = Serial.read();
    switch(sinch) {
      case '0': // 0 key
      case 'u': // u key
        servo1.write(90); // Return to zero position
        servo2.write(90);
        servo3.write(90);
        servo4.write(90);
        servo5.write(90);
        servo6.write(90);
        servo7.write(90);
        servo8.write(90);
        servo9.write(90);
        servo10.write(90);
        servo11.write(90);
        servo12.write(90);
        servo13.write(90);
        servo14.write(90);
        servo15.write(90);
        servo16.write(90);
        servo17.write(90);
        delay(twait*2);
        break;
      // Test out the servos 4 at a time
      case '1': //1 Key - Ankles
        servo1.write(50);
        servo4.write(50);
        servo7.write(50);
        servo10.write(50);
```

```
      delay(twait*2);
      servo1.write(130);
      servo4.write(130);
      servo7.write(130);
      servo10.write(130);
      delay(twait*2);
      break;
    case '2': //2 Key - Knees
      servo2.write(50);
      servo5.write(50);
      servo8.write(50);
      servo11.write(50);
      delay(twait*2);
      servo2.write(130);
      servo5.write(130);
      servo8.write(130);
      servo11.write(130);
      delay(twait*2);
      break;
    case '3': //3 Key - Hips
      servo3.write(50);
      servo6.write(50);
      servo9.write(50);
      servo12.write(50);
      delay(twait*2);
      servo3.write(130);
      servo6.write(130);
      servo9.write(130);
      servo12.write(130);
      delay(twait*2);
      break;
    case '4': //4 Key - Head
      servo13.write(50);
      delay(twait*2);
      servo13.write(130);
      delay(twait*2);
      break;
    case '5': //5 Key - Tail
    case 't': //t Key - Tail
      servo14.write(60);
      servo15.write(120);
```

```
        servo16.write(60);
        servo17.write(120);
        delay(twait*4); // leave time for tail to move.
        servo14.write(120);
        servo15.write(60);
        servo16.write(120);
        servo17.write(60);
        delay(twait*4);
        break;
  // Sequential operations
  case 'd': //d = squat down
        servo1.write(125);
        servo2.write(00);
        servo3.write(45);
        servo4.write(55);
        servo5.write(180);
        servo6.write(135);
        servo7.write(125);
        servo8.write(10);
        servo9.write(40);
        servo10.write(55);
        servo11.write(180);
        servo12.write(140);
        servo13.write(90);
        servo16.write(90);
        delay(twait*2);
      break;
  case 's': // Sit - Sit and shake head
    for (int i=0; i<12; i++){
      servo1.write(sit[0][i]);
      servo2.write(sit[1][i]);
      servo3.write(sit[2][i]);
      servo4.write(sit[3][i]);
      servo5.write(sit[4][i]);
      servo6.write(sit[5][i]);
      servo7.write(sit[6][i]);
      servo8.write(sit[7][i]);
      servo9.write(sit[8][i]);
      servo10.write(sit[9][i]);
      servo11.write(sit[10][i]);
      servo12.write(sit[11][i]);
```

```
        servo13.write(sit[12][i]);
// Wagging tail might tip him over, move tail instead!
        servo16.write(sit[13][i]);
        delay(twait*2);
      }
      break;
    case 'f': // forward key
      for (int i=0; i<12; i++){
        servo1.write(walkf[0][i]);
        servo2.write(walkf[1][i]);
        servo3.write(walkf[2][i]);
        servo4.write(walkf[3][i]);
        servo5.write(walkf[4][i]);
        servo6.write(walkf[5][i]);
        servo7.write(walkf[6][i]);
        servo8.write(walkf[7][i]);
        servo9.write(walkf[8][i]);
        servo10.write(walkf[9][i]);
        servo11.write(walkf[10][i]);
        servo12.write(walkf[11][i]);
        servo14.write(walkf[12][i]);
        servo15.write(walkf[13][i]);
        servo16.write(walkf[14][i]);
        servo17.write(walkf[15][i]);
        delay(twait);
      }
      break;
    case 'b':  // back key
      for (int i=0; i<12; i++){
        servo1.write(walkb[0][i]);
        servo2.write(walkb[1][i]);
        servo3.write(walkb[2][i]);
        servo4.write(walkb[3][i]);
        servo5.write(walkb[4][i]);
        servo6.write(walkb[5][i]);
        servo7.write(walkb[6][i]);
        servo8.write(walkb[7][i]);
        servo9.write(walkb[8][i]);
        servo10.write(walkb[9][i]);
        servo11.write(walkb[10][i]);
        servo12.write(walkb[11][i]);
```

```
        servo14.write(walkb[12][i]);
        servo15.write(walkb[13][i]);
        servo16.write(walkb[14][i]);
        servo17.write(walkb[15][i]);
        delay(twait);
      }
     break;
   case 'r': // right key;
     for (int i=0; i<12; i++){
        servo1.write(turnr[0][i]);
        servo2.write(turnr[1][i]);
        servo3.write(turnr[2][i]);
        servo4.write(turnr[3][i]);
        servo5.write(turnr[4][i]);
        servo6.write(turnr[5][i]);
        servo7.write(turnr[5][i]);
        servo8.write(turnr[6][i]);
        servo9.write(turnr[8][i]);
        servo10.write(turnr[9][i]);
        servo11.write(turnr[10][i]);
        servo12.write(turnr[11][i]);
        delay(twait);
      }
     break;
   case 'l': // left key;
     for (int i=0; i<12; i++){
        servo1.write(turnl[0][i]);
        servo2.write(turnl[1][i]);
        servo3.write(turnl[2][i]);
        servo4.write(turnl[3][i]);
        servo5.write(turnl[4][i]);
        servo6.write(turnl[5][i]);
        servo7.write(turnl[6][i]);
        servo8.write(turnl[7][i]);
        servo9.write(turnl[8][i]);
        servo10.write(turnl[9][i]);
        servo11.write(turnl[10][i]);
        servo12.write(turnl[11][i]);
        delay(twait);
      }
     break;
```

```
        }}
    else{ // nothing to do so gently wiggle tail and mouth.
        servo13.write(85);
        servo16.write(95);
        servo17.write(95);
        delay(twait*10);
        servo13.write(95);
        servo16.write(85);
        servo17.write(85);
        delay(twait*10);
    }
}  // End of program
```

Chapter 10

Walking Hexapod/Spider

18 DOF

One of my objectives in making the spider robot was to only use the same standard parts that were used for the previous robots. Basically my hope was that the spider would all be made with standard parts. However the lower legs would take three straight brackets each for a total of 18! No one seems to be selling just the lower legs by themselves for the spider either. I decided to make the lower legs myself out of Plexiglas.

To make the design for the lower legs first I obtained pictures of them. The pictures were enhanced with edge detection then inverted to be black on white. Then they were enlarged until they matched the mounting spacing of the servos. Then they were printed out and cut out to use as a model. I settled on the simple leg design that is shown below. The leg should be enlarged that so it is 6.25 inches long. This length will result in the spider being lifted 1/2 inch off the ground when its legs are fully retracted. You will need six of them.

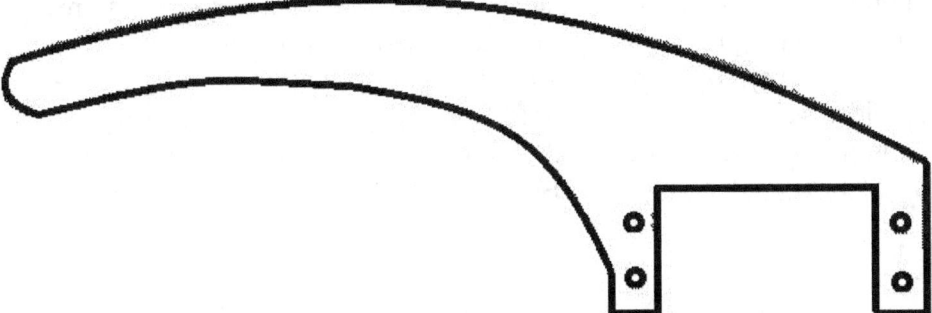

At first the legs were made out of clear 1/4 inch thick plastic cut out with a jig saw. Then the rough edges were filed off. Then they were sanded and painted black. Here is what some of the lower legs looked like when they were done.

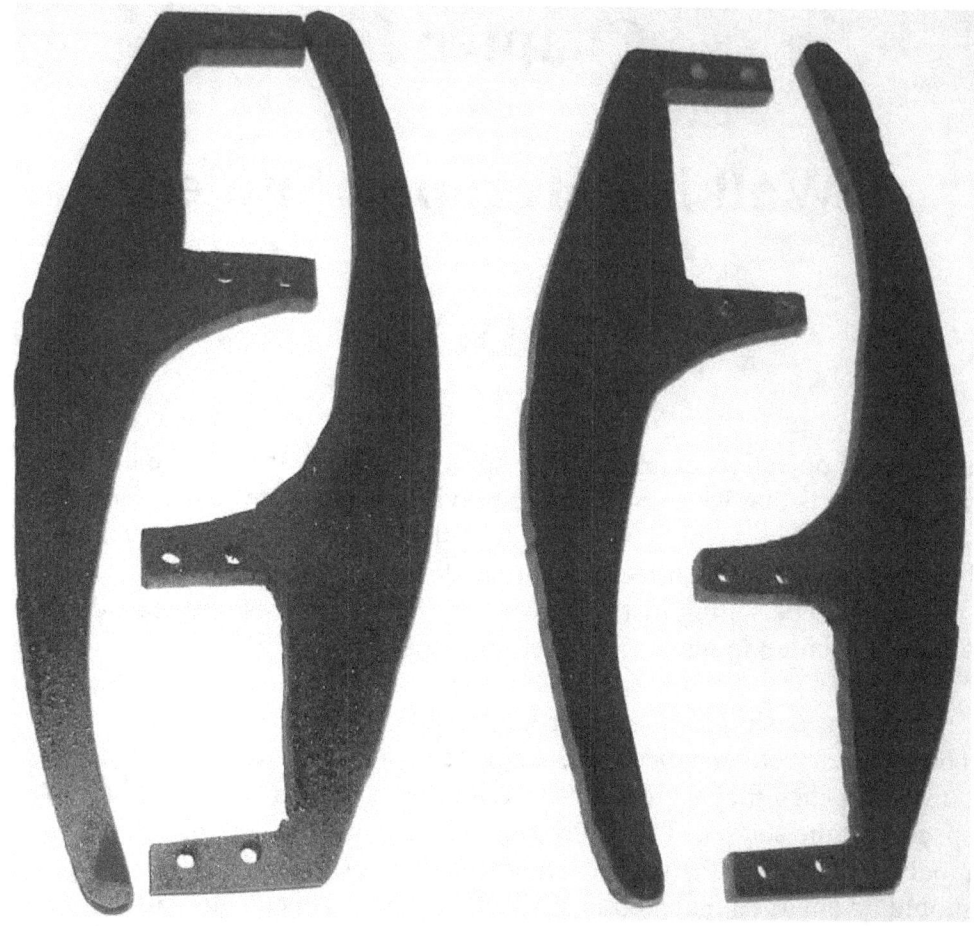

The normal spider body consists of a top and a bottom plate with holes for mounting six servo brackets in between them. Then some spacers are used to keep the plates the right distance apart. I also enhanced and printed out a picture of the body of the spider. That was used to get a rough idea of how to make the spiders body. After some tinkering and rearranging the parts I found a solution. This next picture is the resulting design of the spider body that uses only standard servo bracket parts.

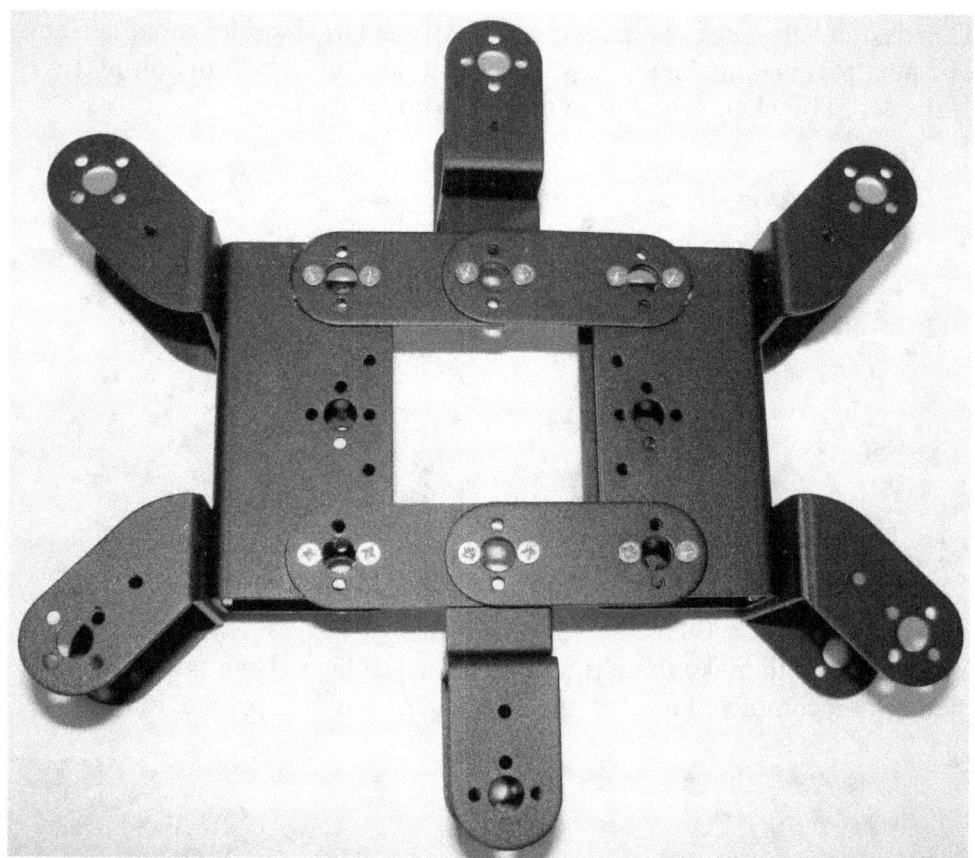

This is the 18 DOF Spider robot parts list:

- 18 x Multi-functional servo mounting bracket
- 18 x MG995 or MG996 Servos
- 8 x Short U-type servo bracket
- 6 x Long U-type servo bracket
- 4 x Angled U-type servo bracket
- 4 x Flat servo bracket
- 2 x L-type servo bracket
- 2 x Robot waist bracket
- 18 x Miniature Ball Radial Bearing
- 18 x Metal Servo Horn Flange
- Lots of screws and nuts
- 6 Spider lower legs
- Optional 2 LED eyes.
-Arduino Uno and Servo Shield

This design very closely replicates some of the larger spider robot designs. The next part is to make the legs. You will need to start with bolting two of the servo brackets together to make the upper leg joint.

You will need to make three right and three left legs. Here is a close up picture of a completed left leg. The right legs are the opposite in design.

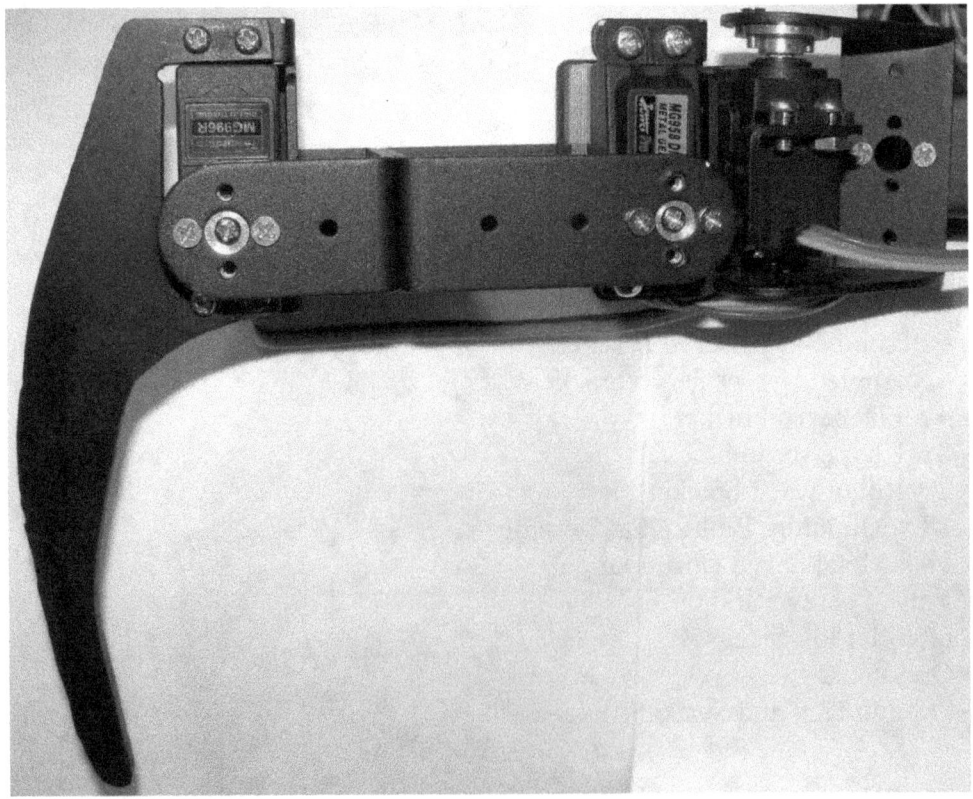

The leg in the last picture is shown has the servos in their 90 degree or "home" position. The top servo should align at 90 degrees with its bracket. In the next picture you can see the 90 degree "home" positions of the top servos.

Once the legs are assembled and attached, the next step is to add the Arduino and then connect the servos. There are mounting holes in the waist bracket that will align with two of the mounting holes towards the back of the Arduino. You will need some 1/2 inch plastic spacers. Here is a picture with how I connected the servos to the shield. The servos are wired up in a circle around the Arduino.

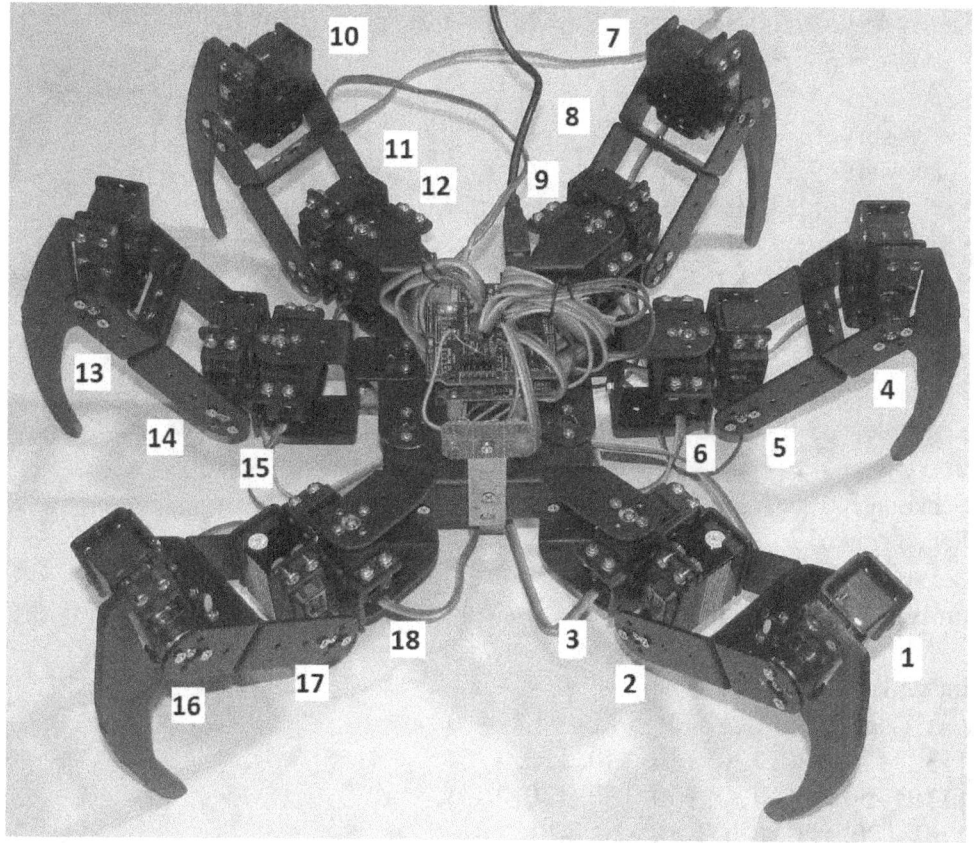

When I first powered up the Spider it had problems getting off the ground. The servos were not strong enough to lift that much weight! Eventually I replaced two of the middle servos with MG958's. With just two of these more powerful servos installed he had no problem getting up into the standing position.

Here is the code to make it walk and turn as well as to test out the servos.

```
// Spider 18 DOF version 2
// November 2015 by Bob Davis
// Mapped as D2 is servo1
// Mapped as D3 is servo2
// Etc.

#include <Servo.h>
// Define our Servo's
Servo servo1; // LF Lower Leg
Servo servo2; // LF Middle Leg
Servo servo3; // LF hip
Servo servo4; // LM Lower leg
Servo servo5; // LM Middle Leg
Servo servo6; // LM Hip
Servo servo7; // LR Lower Leg
Servo servo8; // LR Middle Leg
Servo servo9; // LR hip
Servo servo10; // RR Lower leg
Servo servo11; // RR Middle leg
Servo servo12; // RR hip
Servo servo13; // RM Lower Leg
Servo servo14; // RM Middle leg
Servo servo15; // RM hip
Servo servo16; // RF Lower leg
Servo servo17; // RF Middle Leg
Servo servo18; // RF hip
int twait=100; // Sets time delay between steps

int walkf[18][8] = {
// 0  Leg1 Leg6 Leg2 Leg5 Leg3 Leg4  0
{130, 110, 130, 130, 130, 130, 130, 130},
{130, 150, 130, 130, 130, 130, 130, 130},
{ 90, 120, 120, 120, 120, 120, 120,  90},
{130, 130, 130, 110, 130, 130, 130, 130},
{130, 130, 130, 150, 130, 130, 130, 130},
{ 90,  60,  60, 120, 120, 120, 120,  90},
{130, 130, 130, 130, 130, 110, 130, 130},
{130, 130, 130, 130, 130, 150, 130, 130},
{ 90,  60,  60,  60,  60, 120, 120,  90},
```

```
{ 50,  50,  50,  50,  50,  50,  70,  50},
{ 50,  50,  50,  50,  50,  50,  30,  50},
{ 90, 120, 120, 120, 120, 120,  60,  90},
{ 50,  50,  50,  50,  70,  50,  50,  50},
{ 50,  50,  50,  50,  30,  50,  50,  50},
{ 90, 120, 120, 120,  60,  60,  60,  90},
{ 50,  50,  70,  50,  50,  50,  50,  50},
{ 50,  50,  30,  50,  50,  50,  50,  50},
{ 90, 120,  60,  60,  60,  60,  60,  90},
};

int walkb[18][8] = {
// 0  Leg1 Leg6 Leg2 Leg5 Leg3 Leg4  0
{130, 110, 130, 130, 130, 130, 130, 130},
{130, 150, 130, 130, 130, 130, 130, 130},
{ 90,  60,  60,  60,  60,  60,  60,  90},
{130, 130, 130, 110, 130, 130, 130, 130},
{130, 130, 130, 150, 130, 130, 130, 130},
{ 90, 120, 120,  60,  60,  60,  60,  90},
{130, 130, 130, 130, 130, 110, 130, 130},
{130, 130, 130, 130, 130, 150, 130, 130},
{ 90, 120, 120, 120, 120,  60,  60,  90},
{ 50,  50,  50,  50,  50,  50,  70,  50},
{ 50,  50,  50,  50,  50,  50,  30,  50},
{ 90,  60,  60,  60,  60,  60, 120,  90},
{ 50,  50,  50,  50,  70,  50,  50,  50},
{ 50,  50,  50,  50,  30,  50,  50,  50},
{ 90,  60,  60,  60, 120, 120, 120,  90},
{ 50,  50,  70,  50,  50,  50,  50,  50},
{ 50,  50,  30,  50,  50,  50,  50,  50},
{ 90,  60, 120, 120, 120, 120, 120,  90},
};

int turnr[18][8] = {
// 0  Leg1 Leg6 Leg2 Leg5 Leg3 Leg4  0
{130, 110, 130, 130, 130, 130, 130, 130},
{130, 150, 130, 130, 130, 130, 130, 130},
{ 90, 120, 120, 120, 120, 120, 120,  90},
{130, 130, 130, 110, 130, 130, 130, 130},
{130, 130, 130, 150, 130, 130, 130, 130},
{ 90,  60,  60, 120, 120, 120, 120,  90},
```

```cpp
{130, 130, 130, 130, 130, 110, 130, 130},
{130, 130, 130, 130, 130, 150, 130, 130},
{ 90,  60,  60,  60,  60, 120, 120,  90},
{ 50,  50,  50,  50,  50,  50,  70,  50},
{ 50,  50,  50,  50,  50,  50,  30,  50},
{ 90,  60,  60,  60,  60,  60, 120,  90},
{ 50,  50,  50,  50,  70,  50,  50,  50},
{ 50,  50,  50,  50,  30,  50,  50,  50},
{ 90,  60,  60,  60, 120, 120, 120,  90},
{ 50,  50,  70,  50,  50,  50,  50,  50},
{ 50,  50,  30,  50,  50,  50,  50,  50},
{ 90,  60, 120, 120, 120, 120, 120,  90},
};

int turnl[18][8] = {
// 0  Leg1 Leg6 Leg2 Leg5 Leg3 Leg4  0
{130, 110, 130, 130, 130, 130, 130, 130},
{130, 150, 130, 130, 130, 130, 130, 130},
{ 90,  60,  60,  60,  60,  60,  60,  90},
{130, 130, 130, 110, 130, 130, 130, 130},
{130, 130, 130, 150, 130, 130, 130, 130},
{ 90, 120, 120,  60,  60,  60,  60,  90},
{130, 130, 130, 130, 130, 110, 130, 130},
{130, 130, 130, 130, 130, 150, 130, 130},
{ 90, 120, 120, 120, 120,  60,  60,  90},
{ 50,  50,  50,  50,  50,  50,  70,  50},
{ 50,  50,  50,  50,  50,  50,  30,  50},
{ 90, 120, 120, 120, 120, 120,  60,  90},
{ 50,  50,  50,  50,  70,  50,  50,  50},
{ 50,  50,  50,  50,  30,  50,  50,  50},
{ 90, 120, 120, 120,  60,  60,  60,  90},
{ 50,  50,  70,  50,  50,  50,  50,  50},
{ 50,  50,  30,  50,  50,  50,  50,  50},
{ 90, 120,  60,  60,  60,  60,  60,  90},
};

void setup() {
  servo1.attach(2); // servo on digital pin 2
  servo2.attach(3); // servo on digital pin 3
  servo3.attach(4); // servo on digital pin 4
  servo4.attach(5); // servo on digital pin 5
```

```
servo5.attach(6); // servo on digital pin 6
servo6.attach(7); // servo on digital pin 7
servo7.attach(8); // servo on digital pin 8
servo8.attach(9); // servo on digital pin 9
servo9.attach(10); // servo on digital pin 10
servo10.attach(11); // servo on digital pin 11
servo11.attach(12); // servo on digital pin 12
servo12.attach(13); // servo on digital pin 13
servo13.attach(14); // servo on digital pin 14-A0
servo14.attach(15); // servo on digital pin 15-A1
servo15.attach(16); // servo on digital pin 16-A2
servo16.attach(17); // servo on digital pin 17-A3
servo17.attach(18); // servo on digital pin 18-A4
servo18.attach(19); // servo on digital pin 19-A5
Serial.begin(9600);
Serial.print("Ready");
}

void loop(){
  if ( Serial.available()) {
    char sinch = Serial.read();
    // Test out the servos 6 at a time
    switch(sinch) {
      case 'u': // rise up to 90 degrees
        servo1.write(90); servo2.write(90); servo3.write(90);
        servo4.write(90); servo5.write(90); servo6.write(90);
        servo7.write(90); servo8.write(90); servo9.write(90);
        servo10.write(90); servo11.write(90); servo12.write(90);
        servo13.write(90); servo14.write(90); servo15.write(90);
        servo16.write(90); servo17.write(90); servo18.write(90);
        break;
      case 'd': // Return to down position
        servo1.write(130); servo2.write(130); servo3.write(90);
        servo4.write(130); servo5.write(130); servo6.write(90);
        servo7.write(130); servo8.write(130); servo9.write(90);
        servo10.write(50); servo11.write(50); servo12.write(90);
        servo13.write(50); servo14.write(50); servo15.write(90);
        servo16.write(50); servo17.write(50); servo18.write(90);
        break;
      case '1': //1 Key - Ankles
        servo1.write(60); servo4.write(60); servo7.write(60);
```

```cpp
      servo10.write(120); servo13.write(120); servo16.write(120);
      delay(twait*4);
      servo1.write(120); servo4.write(120); servo7.write(120);
      servo10.write(60); servo13.write(60); servo16.write(60);
      delay(twait*4);
      break;
    case '2': //2 Key - Knees
      servo2.write(70); servo5.write(70); servo8.write(70);
      servo11.write(110); servo14.write(110); servo17.write(110);
      delay(twait*4);
      servo2.write(110); servo5.write(110); servo8.write(110);
      servo11.write(70); servo14.write(70); servo17.write(70);
      delay(twait*4);
      break;
    case '3': //3 Key - Hips
      servo3.write(60); servo6.write(60); servo9.write(60);
      servo12.write(120); servo15.write(120); servo18.write(120);
      delay(twait*4);
      servo3.write(120); servo6.write(120); servo9.write(120);
      servo12.write(60); servo15.write(60); servo18.write(60);
      delay(twait*4);
      break;
    // Sequential operations
    case 'f': // forward key
      for (int i=0; i<8; i++){
        servo1.write(walkf[0][i]);
        servo2.write(walkf[1][i]);
        servo3.write(walkf[2][i]);
        servo4.write(walkf[3][i]);
        servo5.write(walkf[4][i]);
        servo6.write(walkf[5][i]);
        servo7.write(walkf[6][i]);
        servo8.write(walkf[7][i]);
        servo9.write(walkf[8][i]);
        servo10.write(walkf[9][i]);
        servo11.write(walkf[10][i]);
        servo12.write(walkf[11][i]);
        servo13.write(walkf[12][i]);
        servo14.write(walkf[13][i]);
        servo15.write(walkf[14][i]);
        servo16.write(walkf[15][i]);
```

```
      servo17.write(walkf[16][i]);
      servo18.write(walkf[17][i]);
      delay(twait*2);
    }
    break;
  case 'b':  // back key
    for (int i=0; i<8; i++){
      servo1.write(walkb[0][i]);
      servo2.write(walkb[1][i]);
      servo3.write(walkb[2][i]);
      servo4.write(walkb[3][i]);
      servo5.write(walkb[4][i]);
      servo6.write(walkb[5][i]);
      servo7.write(walkb[6][i]);
      servo8.write(walkb[7][i]);
      servo9.write(walkb[8][i]);
      servo10.write(walkb[9][i]);
      servo11.write(walkb[10][i]);
      servo12.write(walkb[11][i]);
      servo13.write(walkb[12][i]);
      servo14.write(walkb[13][i]);
      servo15.write(walkb[14][i]);
      servo16.write(walkb[15][i]);
      servo17.write(walkb[16][i]);
      servo18.write(walkb[17][i]);
      delay(twait*2);
    }
    break;
  case 'l': // left key;
    for (int i=0; i<8; i++){
      servo1.write(turnl[0][i]);
      servo2.write(turnl[1][i]);
      servo3.write(turnl[2][i]);
      servo4.write(turnl[3][i]);
      servo5.write(turnl[4][i]);
      servo6.write(turnl[5][i]);
      servo7.write(turnl[6][i]);
      servo8.write(turnl[7][i]);
      servo9.write(turnl[8][i]);
      servo10.write(turnl[9][i]);
      servo11.write(turnl[10][i]);
```

```
          servo12.write(turnl[11][i]);
          servo13.write(turnl[12][i]);
          servo14.write(turnl[13][i]);
          servo15.write(turnl[14][i]);
          servo16.write(turnl[15][i]);
          servo17.write(turnl[16][i]);
          servo18.write(turnl[17][i]);
          delay(twait*2);
        }
      break;
    case 'r': // right key;
      for (int i=0; i<8; i++){
        servo1.write(turnr[0][i]);
        servo2.write(turnr[1][i]);
        servo3.write(turnr[2][i]);
        servo4.write(turnr[3][i]);
        servo5.write(turnr[4][i]);
        servo6.write(turnr[5][i]);
        servo7.write(turnr[6][i]);
        servo8.write(turnr[7][i]);
        servo9.write(turnr[8][i]);
        servo10.write(turnr[9][i]);
        servo11.write(turnr[10][i]);
        servo12.write(turnr[11][i]);
        servo13.write(turnr[12][i]);
        servo14.write(turnr[13][i]);
        servo15.write(turnr[14][i]);
        servo16.write(turnr[15][i]);
        servo17.write(turnr[16][i]);
        servo18.write(turnr[17][i]);
        delay(twait*2);
      }
      break;
    }
  }
} // End of program
```

Bibliography

Programming Arduino
Getting Started With Sketches
By Simon Mark
Copyright 2012 by the McGraw-Hill Companies

This book gives a thorough explanation of the programming code for the Arduino. However the projects in the book are very basic. It does cover LCD's and Ethernet adapters.

Getting Started with Arduino
By Massimo Banzi
Copyright 2011 Massimo Banzi

This author is a co-founder of the Arduino. This book has a quick reference to the programming code and some simple projects.

Arduino Cookbook
by Michael Margolis
Copyright © 2011 Michael Margolis and Nicholas Weldin. All rights reserved.
Printed in the United States of America.
Published by O'Reilly Media, Inc., 1005 Gravenstein Highway North, Sebastopol, CA.

This book has lots of great projects, with a very good explanation for every project.

Practical Arduino: Cool Projects for Open Source Hardware
by Jonathan Oxer and Hugh Blemings
Copyright © 2009 by Jonathan Oxer and Hugh Blemings